Springer Oceanography

The Springer Oceanography series seeks to publish a broad portfolio of scientific books, aiming at researchers, students, and everyone interested in marine sciences. The series includes peer-reviewed monographs, edited volumes, textbooks, and conference proceedings. It covers the entire area of oceanography including, but not limited to, Coastal Sciences, Biological/Chemical/Geological/Physical Oceanography, Paleoceanography, and related subjects.

More information about this series at http://www.springer.com/series/10175

Chongwei Zheng · Hui Song · Fang Liang ·
Yi-peng Jin · Dong-yu Wang · Yu-chi Tian

21st Century Maritime Silk Road: Wind Energy Resource Evaluation

Springer

Chongwei Zheng
Dalian Naval Academy
Dalian, China

Hui Song
Dalian Naval Academy
Dalian, China

Fang Liang
National Security College
National Defense University of PLA
Beijing, China

Yi-peng Jin
The Aerospace City School of RDFZ
Beijing, China

Dong-yu Wang
Wanglan Institute of Marine Technology
Zhuhai, China

Yu-chi Tian
Fudan University
Shanghai, China

ISSN 2365-7677   ISSN 2365-7685   (electronic)
Springer Oceanography
ISBN 978-981-16-4113-8   ISBN 978-981-16-4111-4   (eBook)
https://doi.org/10.1007/978-981-16-4111-4

This Springer imprint is published by the registered company Springer Nature Singapore Pte Ltd.
The registered company address is: 152 Beach Road, #21-01/04 Gateway East, Singapore 189721, Singapore

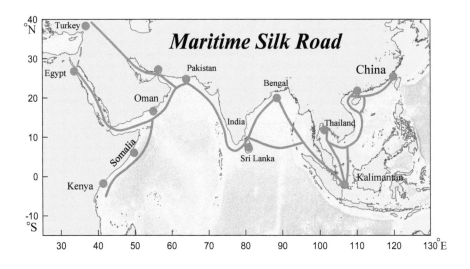

Preface

As the energy and environment crises are accelerating, the smog and the deteriorating ecological environment have been gaining more and more attention nowadays. How to solve these crises has become the shared responsibility for all mankind. Today, conventional energy, such as coal and gas, is in severe shortage. Thus, marine renewable energy (such as offshore wind energy resource) will become the pillar to human society in the twenty-first century and support to our sustainable development. It will also be a new highlight for the Maritime Silk Road, and one of the best solutions to climate change and conventional energy shortage, as well as a good opportunity for international exchange and cooperation. At the same time, the marine renewable energy will also be an important support to realize the peak carbon dioxide emissions and achieve carbon neutrality. Most developed countries have passed laws and regulations such as tax reduction to encourage the development of marine new energy.

Offshore wind energy for years has been under the spotlight, as it eclipses others in terms of being safe, clean, renewable, abundant, widely distributed, as well as saving onshore space. Wind energy is very competitive considering the cost of the power generation and pollution management, as the industry is advancing with the rapid progress in technologies. Using wind power to generate electricity is the main method to exploit this new energy. Besides that, it is also applied in desalination, navigation, and wind heating. In coastal areas, high GDP has also brought high pressure for electricity supply. Thus, by exploiting the offshore wind energy according to local conditions, local authorities could effectively respond to the energy crisis and promote sustainable development of local communities. As for remote islands and reefs, which are in desperate need of electricity, utilizing marine resources and exploiting offshore wind energy could not only fill the gap of power supply, but also protect the ecological environment, avoiding the pollution caused by diesel engines.

In most cases, remote islands are an important support for deep-sea exploration. As these islands are far away from the continent, generating electricity under such a condition has always been a global challenge. Diesel, delivered by ship, has served as a common solution for remote islands to deal with the energy crisis, but it is not without its problems: The supply lines are too long and could easily be subjected to extreme weather; the ecological environment on such islands is too fragile to be restored once polluted by diesel engines. Thus, offshore wind energy and desalination programs could be an antidote to the power shortage and a blessing to island dwellers and countries along the Maritime Silk Road which are facing energy and environment crises.

Currently, onshore wind power generation technologies are relatively mature. However, when it comes to offshore, only few countries are equipped with mature technologies, while wind power generation in most countries and regions is still in its infancy. Besides, wind power varies hugely in different regions and seasons. Therefore, one of the basic principles for massive wind power development is "evaluation of resources and planning go first." Based on solid and detailed wind power evaluations and planning which covers its development and grid construction, we could well manage the wind power and exploit it effectively.

This book aims to establish a wind energy evaluation system, to provide scientific research for site selection, daily operation, and long-term planning of wind power generation. Firstly, we analyze the advantages and disadvantages of offshore wind power, then further discuss the status quo and challenges for wind power programs along the Maritime Silk Road, and offer suggestions. A wind energy evaluation system was proposed with the Maritime Silk Road as a case study, including climatic features of wind power (temporal–spatial distribution), long-term climatic trend and mechanism, short-term forecast of wind energy, mid- and long-term projection of wind energy, technology of wind energy evaluation on key point or vital region, and offshore wind energy dataset construction, to provide systematic and scientific reference for wind power evaluation and utilization. We hope the research could make contribution to easing the energy and environment tension, promoting the living standards of people along the Maritime Silk Road and break the shackles of power shortage.

Dalian, China   Chongwei Zheng
Dalian, China   Hui Song
Beijing, China   Fang Liang
Beijing, China   Yi-peng Jin
Zhuhai, China   Dong-yu Wang
Shanghai, China   Yu-chi Tian

Acknowledgements

First of all, we would like to thank academician Chongyin Li for providing the excellent guidance on our academic career. On the voyage of life, you have kindled the light of hope for us. What you have done enriches our mind and broadens our view. The love and care you have given us will encourage us to go through a long and arduous journey. We honor you sincerely.

This work was supported by the open fund project of Shandong Provincial Key Laboratory of Ocean Engineering, Ocean University of China (kloe201901), the open fund project of State Key Laboratory of Numerical Modeling for Atmospheric Sciences and Geophysical Fluid Dynamics, Institute of Atmospheric Physics, the Chinese Academy of Sciences, the open research fund of State Key Laboratory of Estuarine and Coastal Research (No. SKLEC-KF201707), the Major International (Regional) Joint Research Project of National Science Foundation of China (No. 41520104008), and the National Key R&D Program of China (No. 2018YFA0605604; No. 2017YFC1501802). All the authors would like to thank ECMWF for providing the ERA-Interim data and IPCC for providing the CMIP data.

Series Publications on the 21st Century Maritime Silk Road

Contents

About the Authors

Dr. Chongwei Zheng obtained his Ph.D. degree in Atmospheric Sciences from National University of Defense Technology. He is the creator of Marine Resources and Environment Research Group on the Maritime Silk Road. His research fields cover physical oceanography, marine new energy evaluation, wind and wave climate, climate change, air–sea interaction, and so on. He has published more than 90 papers in many peer-reviewed journals as the first author or corresponding author, including 38 papers indexed by SCI and 15 papers indexed by EI. He has published six books as first author. He is a reviewer for more than 40 high-impact journals (including Renewable and Sustainable Energy Reviews, Applied Energy, Remote Sensing of the Environment, etc.). He is also a monograph reviewer for Elsevier. In 2017, his project titled "Wave energy and offshore wind energy resources evaluation of the Maritime Silk Road" won the Marine Engineering Science and Technology award from the China Association of Oceanic Engineering.

Prof. Hui Song was born in Laiwu city, Shandong Province, in 1969. He got the master of engineering in Navigation Technology from Dalian Naval Academy in 1996. He has studied abroad at the Collège Interarmées de Défense (CID). His research interests include the navigation technology, education and teaching methods, innovative talent training, and so on. He is mainly engaged in the teaching research and teaching management in the field of navigation. In recent years, he has won the first prize of Chinese National Teaching Achievement Award. He has published six academic research papers on navigation and education management. He has also published one academic book on navigation and education management.

Prof. Fang Liang is doctor and doctoral supervisor at the National Security College of China National Defense University. She graduated from the National Defense University with a major in Operations Research. In 2015, she was awarded the honorary title of National March 8th Red-Banner Pacesetter, a director of the Chinese Ocean Society, and the Deputy Secretary General of a discipline Professional Committee of Oceanographic Society of China. She has long been engaged in teaching and research in the fields of ocean planning and design. He has published many books such as the academic monograph "On Important Maritime Channels." She published more than 60 academic articles. In 2020, she won the national second prize of the National Vocational College Teaching Ability Competition.

Ms. Yi-peng Jin graduated in 2019 with a master degree in foreign linguistics and applied linguistics from the Graduate School of Translation and Interpretation (GSTI) in Beijing Foreign Studies University and is a translator currently based in Beijing. She is now working at the RDFZ Aeroplane School as an English teacher. Her research field covers major areas in translation and simultaneous interpretation. She also received her bachelor degree in BFSU in international relations and diplomacy. She is qualified as a CATTI Level II interpreter and has won the second place in 2019 GSTI Interpretation Competition. During her time at school and after graduation, she's been active in various fields of translation and interpretation. She served as the simultaneous interpreter for the 2019 China International Fair for Trade in Services (CIFTIS) and has also provided interpretation and translation for United Nations Development Program (UNDP), Federation of University Sports of China, Destination Canada, RDFZ, and many other institutions.

Mr. Dong-yu Wang , 41 years old, is Director of the Wanglan Marine Technology Institute. He has a Master's degree in Ocean Engineering from South China Sea Institute of Oceanology, Chinese Academy of Sciences. During his work at the South China Sea Institute of Oceanology, Chinese Academy of Sciences, Wang served as an assistant researcher in the research group of Dr. Zhang Jingxiang, a senior engineer of marine buoys and submerged buoy research and development, and participated in Dr. Zhang's several projects including large-scale deep-sea buoy, the preliminary investigation buoy and mooring system of the Hong Kong-Zhuhai-Macao Bridge, the South China Sea Petroleum Development Zone survey buoy subject, etc. He won two second prizes and three third prizes for scientific and technological progress by Chinese Academy of Sciences and Guangdong Academy of Sciences, respectively. He obtained several invention patents, including the double-buoy system, small wave generator, high-efficiency wind power generation, and more than 20 utility model patents.

Dr. Yu-chi Tian was born in 1990, and he got a double master degree in biomedical engineering from and Northeastern University in China and University of Dundee in UK, respectively. And he is studying for a doctorate degree in biomedical engineering at Fudan University now. His research field is focused on artificial intelligence and medical treatment and health. He has obtained the "Dean Scholarship" in University of Dundee, UK, in 2016, and has been awarded the Award for Outstanding Ph.D. student as Fudan University for the academic year 2019–2020. He has hold on a project of campus resource linkage platform funded by Fudan

Fanhai Venture Foundation. Also, his research project of "Prediction and evaluation of the effect of HCC immunotherapy based on MRI imaging" won funding from the Yunfeng Foundation.

Chapter 1
Introduction

A community with a shared future for mankind has been advocated by China with pragmatic development initiatives, one of which would be the Maritime Silk Road. It is a major step forward, an initiative that could not only benefit people along the Road, also a blue belt that links the Chinese dream with the global one (Zheng et al. 2018a, 2018b). However, there are many difficulties in the construction process. The Maritime Silk Road covers a long route and vast sea areas with complex natural conditions, and outdated infrastructure, lack of oceanographic data and basic studies have all been severely constraining our abilities to research, explore and exploit marine resources. The overall fragile power supply along the Road has been a barrier to the building of the Maritime Silk Road. Generally speaking, the total power consumption along the Belt and Road Initiative is only 61% of the world's average amount (Jiang et al. 2019). The key to effectively building the Maritime Silk Road is to solve the electricity crisis.

The offshore wind energy, large in amount, wide in distribution and available under all conditions, can be applied to power generation and desalination, as well as water pumping, wind-heating and other projections (Soukissian and Papadopoulos 2015; Xydis 2015). And it provides strong power support to the building of the Maritime Silk Road. It is not only one of best choices to break the crisis but also a blessing to marine ecological protection, people living along the Road and the exploration of remote sea areas. It provides a good opportunity for international exchange and cooperation and an antidote to climate change and conventional energy shortage. At the same time, the offshore wind energy will also provide important support for mandkind to achieve the carbon neutrality target. As wind power variables hugely in different regions and seasons, one of the basic principles for massive wind power development is "evaluation of resources and planning go first". Based on solid and detailed wind power evaluations, we could well-manage the wind power and exploit it effectively.

© The Author(s), under exclusive license to Springer Nature Singapore Pte Ltd. 2021   1
C. Zheng et al., *21st Century Maritime Silk Road: Wind Energy Resource Evaluation*,
Springer Oceanography, https://doi.org/10.1007/978-981-16-4111-4_1

1.1 Advantages of Offshore Wind Energy

The deteriorating energy and environment crises have been severely threatening the living and sustainable development of all man-kind. Currently, conventional energy such as coal and oil has becoming more and more scarce, and humans are putting new hopes on new energy which is safe, clean, renewable, abundant and widely-distributed. With the advancing of technologies, wind power will be extremely competitive considering the external cost of power generation and pollution (Junginger et al. 2004; Blanco 2009). Wind energy is mainly used for power generation, as well as desalination, navigation, water lifting, irrigation, heating, etc. Offshore wind energy has the following advantages (Zheng 2011):

(1)   Remote islands and reefs are an important support for marine exploration. However, remote islands and reefs are in desperate need of electricity. Utilizing marine resources and exploiting offshore wind energy could not only fill the gap of power supply, but also protect the ecological environment, avoiding the pollution caused by diesel engines, thus to significantly increase the living conditions and promote their sustainable development as offshore strategic spots. Besides, it can also empower beacons and buoys so as to decrease the relevant cost for maintenance.

(2)   Adequate infrastructure is the prerequisite for commercial development on remote islands such as tourism. Exploiting off shore wind power according to local conditions could not only serve as an antidote to power shortage, but offshore wind farms have become beautiful tourist attractions themselves and could contribute to economic growth. Furthermore, adequate infrastructure could increase the living standards of dwellers and ease the energy and environmental crisis for countries along the Road and in turn promotes its own development.

(3)   Compared with onshore wind power, offshore wind power could generate more electricity: the wind speed 10 km off the coast is 25% higher than that along the coast and suffers less influence. Usable wind power resources are 3 times larger than those onshore (Tambke et al. 2005; Wang et al. 2015; Yao et al. 2007), which promotes better collection and transformation of those resources.

(4)   Offshore wind power saves land resources and the trouble to relocate residence, which would otherwise prove to be a heavy cost. It also minimizes the noise and light pollution and has little influence to human activities.

(5)   The height of the wind turbine towers can be reduced and so does the cost of the turbines, as the ocean surface is less rough with smaller friction and smoother underlying surfaces and higher wind speed at lower altitude.

(6)   The service life of turbines offshore can be prolonged to 50 a in design and their bases can be recycled, as the low turbulence intensity and small friction of the sea surface wind (Li and Yu 2004).

(7)   Technology for wind power generation is rather mature and ready for mass commercial exploration. And the technology would be continually advancing

as some European countries have commercialized the use of offshore wind power since 2001.

(8)   The effective wind speed occurrence (EWSO) is over 60% for most water areas around the world (Zheng et al. 2016). In other words, at least over half a year these water areas are suitable for wind power generation. However, due to the limit of daylight, the useable period of solar power is less than 50 percent of a year.

(9)   There are vast and contiguous areas on the sea, which is more optimistic than that onland.

(10)   With the fast development of oceanic observation technology and computer technology, there are more and more data on marine research which could promote wind power evaluation for vast sea areas. Previous researchers have done tremendous work on wind power evaluation globally and could provide scientific guidance for offshore siting (Zheng and Pan 2012; Zheng et al. 2013a, b). The accuracy of short-term reports on wind farms and wind power resources is growing and could guarantee the running of offshore wind power business (Zheng et al. 2014). The accuracy of mid- and long-term reports on wind farms and wind power resources is also growing which could serve as reference to the mid- and long-term planning of offshore wind power exploitation.

(11)   Previous researchers have also done a lot of work on marine ecological feature analysis for wind power exploitation, so as to guarantee its safety and efficiency (Zheng 2013; Zheng et al. 2014b, 2015, 2013d).

(12)   The bases of offshore turbines could also serve as artificial reefs and provide more food sources for fish. Besides, because the surrounding areas are off limits to boating, fishing and trawling under most circumstances, they have become a refuge for fish and other marine mammals.

1.2   Disadvantages of Offshore Wind Energy

Effect on marine ecology: Most offshore wind farms are built in shallow waters. However, these productive sea areas are inhabitants to all kinds of marine species. The running of offshore wind engines would affect the breeding and living of marine life, and thus such influence should be fully taken into allowance (Yuan et al. 2014; Bergström et al. 2014; Mann and Teilmann 2013; Leung and Yang 2012; Tabassum-Abbasi et al. 2014; Saidur et al. 2011). (1) The noise of wind turbine has an effect on fish's and mammals' the hearing range, communication and sense of directions. When the noise source is audible, it may evoke behavior or other responses in animals. (2) Marine benthos plays a crucial part in maintaining a healthy and stable marine ecological environment. And these organisms may suffer from noise, vibration, temperature change, pollution, electromagnetic and other disturbance while building facilities and submarine cables, running and closing the wind engines and installing their basement. (3) During the construction of offshore wind engine basement, the sound of piling may cause hearing impairment to marine life. (4) Massive offshore wind farms may have an impact on sea surface roughness which affects atmospheric circulation and climate.

Challenges in grid connection: Grid connection of offshore wind farms is faced with two technical challenges: the transmission of offshore wind power and the effect of wind power farm's stability on the grid. There are still technical barriers in the controllability of wind energy and the control technology, such as active pitch control, automatic control and automatic stop. And the same goes to short-term forecast of wind power and collective control strategies for wind farms.

High cost: Offshore wind turbines must be airtight, dry, heat-exchanging and antiseptic to adapt to the terrible ocean environment. Other facilities needed to install and maintain such as boarding platforms and lifting machines as turbines are mostly of high unit capacity. According to statistics, an on-land wind power project takes 8000 yuan RMB per kilowatt. However, the cost could increase to 16,000 to 20,000 yuan RMB per kilowatt for offshore wind power while building and maintenance take up most of it.

1.3   Frame Structure of This Book

This book aims to establish a wind energy evaluation system, to provide scientific research for site selection, daily operation and long-term planning of wind energy development. We first analyze the advantages and disadvantages of offshore wind power, further discusses about the status quo and challenges for wind power programs along the Maritime Silk Road and offer suggestions. Then an offshore wind energy evaluation system is designed. At last, the technology of offshore wind energy evaluation on key islands or vital region is presented.

The book is organized as follows:

Chapter 1 Introduction. The authors mainly discuss the energy crisis and electricity requirement of the Maritime Silk Road, and the advantage and disadvantage of offshore wind energy resources.

Chapter 2 Analysis of current assessment status, difficulties and countermeasures of wind energy evaluation. Firstly, we discussed the research status and difficulties (detail research on the climatic characteristics of wind energy, macro and micro-scale energy classification, relationship between wind energy and key indexes, short-term forecast of wind energy, climatic variation of wind energy, mid-long term projection of wind energy, energy evaluation on key nodes, offshore wind energy dataset construction). Then the countermeasures to deal with these difficulties are provided.

Chapter 3 At home and abroad, this book carries out the first detailed investigation on the climate characteristics of wind energy resource along the Maritime Silk Road. A series of key parameters are defined: availability of wind energy (effective wind speed occurrence, EWSO), available level occurrence (ALO, occurrence of WPD greater than 100 W/m^2)、moderate level occurrence (MLO, occurrence of WPD greater than 150 W/m^2)、rich level occurrence (RLO, occurrence of WPD greater than 200 W/m^2)、 excellent level occurrence (ELO, occurrence of WPD greater than 300 W/m^2)、 superb level occurrence (SLO, occurrence of WPD greater than 400 W/m^2), to describe the availability and richness of wind energy. Systematically considering the wind power density, energy availability, energy level occurrences, energy stability, monthly and seasonal variability of wind energy, energy storage, the temporal-spatial distribution characteristics of offshore wind energy along the Maritime Silk Road are analyzed, to provide theoretical support for the energy classification, short-term forecast and long-term projection of wind energy.

Chapter 4 Analysis of the historical climatic trend of offshore wind energy resource along the Maritime Silk Road. The annual trend, and regional and seasonal difference of the trends of a series of key indicators of wind energy resource along the Maritime Silk Road are calculated and analyzed for the first time, to provide a theoretical foundation for the improvement of medium - and long-term projection capacity for wind energy projection.

Chapter 5 An all-elements short-term forecasting scheme of wind energy resource was proposed, with the Maritime Silk Road as a case study. The traditional forecasting of wind energy mainly includes the wind field or the wind power density. In this book, an all-element short-term forecasting system of offshore wind energy resource was designed, comprehensively including the wind field, wind power density, energy availability, energy level occurrences, energy storage, etc., as well as the forecasting of wind energy of key nodes (hourly wind power density and wind direction, wind energy rose, etc.), to provide reference for the daily operation of wind power generation and so on.

Chapter 6 An all-elements long-term projection scheme of wind energy resource was proposed, with the Maritime Silk Road as a case study. Using CMIP data to carry out the long-term wind energy projection along the Maritime Silk Road for the next 40 years, which includes a series of key indicators, such as: the wind power density, energy availability, energy level occurrences, energy stability, monthly and seasonal variability, energy storage for the next 40 years, to provide scientific reference for the long-term planning of wind energy utilization.

Chapter 7 Technology of offshore wind energy evaluation on key island or vital region, with the Sri Lankan waters as a case study. Systematically considering the wind power density, energy availability, energy level occurrences, energy direction (co-occurrence of wind power density-wind direction), climatic variation of wind energy, wind class, the wind energy of Sri Lankan waters for the future and for the past are analyzed and compared, to provide reference for the long-term planning of wind energy development and technical approach for the mid-long term projection of wind energy resource.

Chapter 8 Construction of temporal-spatial characteristics dataset of offshore wind energy resource. The first open-ended and non-profit dataset of spatial–temporal characteristics of wind energy resource for the Maritime Silk Road at home and abroad was established, to provide data support for offshore wind energy resource evaluation. A wind energy resource, which should include the temporal-spatial characteristics, energy classification, climatic variation of wind energy, long-term projection of wind energy, wind climate, etc., was prospected to be established to provide a systematic data support for power plant site selection, daily operation, long-term planning, environmental safety guarantee for resource development.

References

Bergström L, Kautsky L, Malm T, Rosenberg R, Wahlberg M, Capetillo NA, Wilhelmsson D (2014) Effects of offshore wind farms on marine wildlife—a generalized impact assessment. Environ Res Lett 9:034012 (12). https://doi.org/10.1088/1748-9326/9/3/034012

Blanco MI (2009) The economics of wind energy. Renew Sustain Energy Rev 13(6–7):1372–1382

Jiang Y, Wu MQ, Huang CJ, Niu Z (2000-2019) Collection of data on overseas power projects in the belt and road initiative (2000–2019). China Scientific Data. https://doi.org/10.11922/csdata.2019.0069.zh

Junginger M, Faaij A, Turkenburg WC (2004) Cost reduction prospects for offshore wind farms. Wind Eng 28:97–118

Li XY, Yu Z (2004) Developments of offshore wind power. Acta Energiae Solaris Sinica 25(1):78–84

Leung DYC, Yang Y (2012) Wind energy development and its environmental impact: a review. Renew Sustain Energy Rev 16:1031–1039

Mann J, Teilmann J (2013) Environmental impact of wind energy. Environ Res Lett 8:035001 (3). https://doi.org/10.1088/1748-9326/8/3/035001

Saidur R, Rahim NA, Islam MR, Solangi KH (2011) Environmental impact of wind energy. Renew Sustain Energy Rev 15:2423–2430

Soukissian T, Papadopoulos A (2015) Effects of different wind data sources in offshore wind power assessment. Renewable Energy 77:101–114

Tabassum-Abbasi, Premalatha M, Abbasi T, Abbasi SA (2014) Wind energy: Increasing deployment, rising environmental concerns. Renew Sustain Energy Rev 31:270–288

Tambke J, Lange M, Focken U (2005) Forecasting offshore wind speeds above the North Sea. Wind Energy 8:3–16

Wang JZ, Qin SS, Jin SQ, Wu J (2015) Estimation methods review and analysis of offshore extreme wind speeds and wind energy resources. Renew Sustain Energy Rev 42:26–42

Xydis G (2015) A Wind energy integration analysis using wind resource assessment as a decision tool for promoting sustainable energy utilization in agriculture. J Clean Prod 96:476–485

Yao XJ, Sui HX, Liu YM (2007) The development and current status of offshore wind power technology. Shanghai Electricity 2:111–118

Yuan Z, Ma L, Wang JK (2014) Study on the influence of noise from offshore wind turbines on Marine life. Ocean Develop Manage 31(10):62–66

Zheng CW (2011) Wave energy and other renewable energy resources in South China Sea: advantages and disadvantages. J Subtropical Resour Environ 6(3):76–81

Zheng CW, Li CY, Jing P, Liu MY, Xia LL (2016) An overview of global ocean wind energy resources evaluation. Renew Sustain Energy Rev 53:1240–1251

Zheng CW, Pan J (2012) Wind energy resources assessment in global ocean. J Nat Resour 27(3):364–371

Zheng CW, Jia BK, Guo SP, Zhuang H (2013) Wave energy resource storage assessment in global ocean. Resour Sci 35(8):1611–1616

Zheng CW, Zhuang H, Guo SP, Jia BK (2013) Analyses on wind energy energy resources in East China Sea based on QN wind data. Water Power 39(9):5–8

Zheng CW, Zhou L, Song S, Su Q (2014) Forecasting of the China Sea wind energy density. J Guangdong Ocean Univer 34(1):71–77

Zheng CW (2013) Statistics of gale frequency in global Oceans. J Guangdong Ocean Univer 33(6):77–81

Zheng CW, You XB, Pan J et al (2014) Feasibility analysis on the wind energy and wave energy resources exploitation in Fishing Islands and Scarborough Shoal. Marine Forecasts 31(1):49–57

Zheng CW, Pan J, Sun W et al (2015) Strategic of the ocean environment of the 21st century maritime silk road. Ocean Develop Manage 32(7):4–9

Zheng CW, Lin G, Shao LT (2013) Frequency of rough Sea and Its long-term trend analysis in the China Sea from 1988 to 2010. J Xiamen Univer (natural Science) 52(3):395–399

Zheng CW, Li CY, Wu HL, Wang M (2018a) In: 21st century maritime silk road: construction of remote Islands and Reefs. Springer

Zheng CW, Xiao ZN, Zhou W, Chen XB, Chen X (2018b) In: 21st century maritime silk road: a peaceful way forward. Springer

Chapter 2
Research Status, Difficulties and Countermeasures of Offshore Wind Energy Evaluation of the Maritime Silk Road

One of the best choices to break energy shackles along the Maritime Silk Road is to evaluate and explore the offshore wind energy. However, such evaluations are still relatively few owing to the difficulties faced by marine observation and shortage of research resources in spite of all the efforts that has been done (Zheng et al. 2018a; 2018b; Zheng et al. 2019c, 2019d; Jiang et al. 2019; Zheng and Li 2018; Zheng et al. 2019b; Zheng 2011; Zheng 2018a, 2018b, 2018c, 2018d; Niu et al. 2019; Dou and Xie 2020; Le et al. 2020). And this has been a barrier to developing offshore wind power into a bigger scale as an industry. Therefore, one of the basic principles for massive wind power development is "evaluation of resources and planning go first". Generally offshore wind power evaluation has gone through four stages (Zheng et al. 2016): (1) Early observation stage. Some researchers groundbreakingly evaluate wind power using equipment like buoys and ship news (Youm et al. 2005). However, such equipment is too scarce to cover all the marine areas. Therefore, it would be impossible to evaluate resources at a large scale. (2) Satellite observation stage. Satellite can evaluate offshore wind power all over the world. Since each satellite takes a relatively long revisit period, multi-satellite combination is required to avoid missing out weather processes so as to increase calculation accuracy, which proves to be another challenge. (3) Numeral simulation stage. Numeral simulation could be used to carry out detailed wind power evaluation and study on marine areas without observation equipment. Yet the simulation results in special terrains are still to be improved (Carvalho et al. 2014). (4) Reanalysis stage. With the rapid development of maritime observation, computer technologies, more and more data are available for wind power evaluation. And the reanalysis data is also widely applied in evaluation due to the advance of variational assimilation (Chadee and Clarke 2014).

Despite all the work and efforts done by previous researchers, generally speaking, studies on wind power evaluation along the Maritime Silk Road are still relatively scarce for now. Besides, though tremendous work has been done on temporal and special distribution of certain aspects of wind power, there are still challenges facing detailed studies on climate features, mapping of energy classification, long-term variation, short-term forecasting and long-term projection. However, these are urgent requirements for precise siting, business operating and mid- and long-term planning throughout resource exploitation. This chapter will present the status quo of wind power evaluation along the Road, analyze challenges that we are facing and provide solutions.

2.1   Researches on Wind Power of the South China Sea

Zheng et al. (2012, 2013) defined three key indicators of energy resources: the occurrence of wave power density at different levels (energy level occurrences), effective wind speed occurrence (EWSO) and effective wave height occurrence (EWHO), and pointed out that these indicators are the major description methods of the richness and the availability of resources. Based on the Cross-Calibrated, Multi-Platform (CCMP) wind data and hindcast wave data, Zheng et al. (2012, 2013) carried out a systematic research on the offshore wind power and wave power in the South China Sea and the East China Sea, and classification of offshore and near-shore wind power, by comprehensively analyzing the wind power density, energy level occurrences, availability, stability, storage and long-term trend of resources. The results are as follows: Across most of the China seas, the average wind power density is above 100 W/m^2 and the strong wind energy are mainly found over the Ryukyu Islands, the Luzon Strait and the Pingshun Island. The distribution of annual average wave power density is in line with that of wind power density at the same time intervals. The general value of wind power density is over 50 W/m^2 and that of wave power density over 2 kW/m with both occurrence exceeding 70% and 50% separately in the East and the South China Sea. Generally, the wind power density is more stable than the wave power density while both are the least stable in July. From 1988 to 2009, the annual wind power density and wave power density exhibit a remarkably upward trend. The South China Sea is abundant in wind resources: the Luzon Strait and its western sea area, the southeastern sea area of Indo-China Peninsula, the middle of Taiwan Strait are all sites with high wind energy; those with moderate to high wind energy are located at the mid of the South China Sea running in a northwest-southeast direction. As for the wave energy in China, the bulk of the East and the South China Sea are sites having moderate wave energy; moderate to high energy sites are mainly distributed in the east of Taiwan Island, Luzon Strait and the mid of northern South Sea; the Yellow Sea and the Bohai Sea are sites that have consistently low wave and wind energy resources; however, there is no site that has high wave energy.

In 2015, based on the CCMP wind data and hindcast wave data, considering a series of key parameters, Zheng and Li (2015a) conducted a systematic research on the wind and wave energy of island in the South China Sea. The results show that this area is abundant of wind and wave energy resources suitable for exploitation. (1) Except for extreme conditions, wind and wave energy can be exploited throughout the whole year with a peak period emerging from December to January of the next year. Monthly average wind power density is around 370 W/m^2, wave power density is around 20 kW/m. Energy is enough to support utility scale projects throughout the year, even in April and May when it's at its lowest. (2) The availability of resources and energy level occurrences present promising figures: the probability of monthly effective wind speed occurrence exceeds 70%; the probability of occurrence of effective wave height, wind power density over 50 W/m^2 and wave power density over 2 kW/m are over 50% for most of the year. (3) In the targeted sea area, waves with a height of 2–3 m and a period of 6–7 s generated most of the wave power, accounting for 14.6%. (4) In that area, wind energy is mainly contributed by ENE, NE, SW WSW wind, while 100–300 W/m^2 wind power density has the highest occurrence probability and that over 1000 W/m^2 is mainly contributed by WSW wind. Wave energy is mainly generated by NNE and WSW swells while 0–5 kW/m and 5–10 kW/m levels have the highest occurrence probability, (5) From 1988 to 2011, wind power density of the targeted sea area shows no trend and wave power density presents an increase trend at a speed of 0.25 (kW/m)/yr. (6) During summer, winter and the transition period from the summer monsoon to the summer monsoon, wind energy and wave energy have a good stability, while in May, the lowest stability. (7) The total storage of wind energy is 2050 (kW·h)/m^2 with 1722 (kW·h)/m^2 effective storage; that of wind energy is 84,079 (kW·h)/m with 66,336 (kW·h)/m effective storage. Zheng and Li (2015b) also conducted a detailed analysis on wind and wave climate of islands in the South Sea, including seasonal features of wind and wave, wind power level occurrence, wave power level occurrence, strong wind direction occurrence and strong wave direction occurrence, extreme wind speed and wave height and long-term variation of wind speed and wave height, so as to provide insurance to resource exploration and offshore installation.

Chen et al. (2017) used 6 buoys to calculate the characteristics of wind and wave power in inshore areas of Shenzhen (Dapeng Bay, Daya Bay, Shenzhen Bay, these area with 3–22 m deep waters) and found that the wind speed of 2.5 m above sea surface is 3.1 to 4.1 m/s; wind power density 37 to 94 W/m^2; significant wave height less than 1 m; wave period 3 to 7 s; wave power density within 1 kW/m. Chen et al. (2017) also pointed out that the targeted sea waters are not suitable for wave power exploration because they are shallow and relatively closed inside the port.

Liu et al. (2018) used the Weather Research and Forecast (WRF) model driven by 30 years of regional meteorological data from ECMWF to conduct numeral stimulation on the wind power among coastal waters of China. Liu et al. found that wind power is abundant in the East China Sea and the South China Sea. Especially, the annual average wind power density reaches 800 W/m^2 at a 90-m height at the Taiwan Strait.

Albani et al. (2018) used 35 years of Climate Forecast System Reanalysis data and 10 years of meteorological wind data to analyze the features of wind energy at Mersing and Kijial on the east coast of peninsular Malaysia and Kudat in Sabah. The research shows that ENSO has significant influence on the changes of wind speed in Malaysia and further impacts the wind energy. Albani et al. pointed out that at Mersing, capacity factor is relatively high during the time for the northeast monsoon (21.32%) and relatively low the southeast monsoon (3.71%); at Kijal, this factor bears the same feature: relatively high for the northeast monsoon (10.66%) and relatively low the southeast monsoon (5.19%); however, at Kudat, capacity factor rises up to its peak during the time for northeast monsoon (36.42%) with that being 24.61% during the winter monsoon.

Jiang et al. (2014) calculated the wind power density at a 10-m height above sea surface to evaluate wind energy resources of the South China Sea. Then the wind power density at a 70-m height was calculated according to the wind profile index. Jiang et al. found that the wind power density, lower in spring and summer whereas higher in autumn and winter, ranges from grade 4 to grade 7 which is suitable for power generation. Among islands and reefs in the South China Sea, the wind power installed capacity of Xisha Islands is the highest, that of Dongsha Islands and Zhongsha Islands both at the third level; and that of Nansha Islands is at the first to second level. Among all the islands and reefs of this area, the wind power installed capacity of the Dongsha Islands, Scarborough Reef of Zhongsha Islands, Discovery Reef, Bombay Reef, Zhaoshu Dao of Xisha Islands and Mariveles Reef, Ladd Reef, Cornwallis South Reef of Nansha Islands is relatively high, and wind power generation should be developed on these islands first.

Chen et al. (2014) used meteorological data of Yongxing weather station in Xisha collected from 1973 to 2011, Weibull statistical distribution and wind power curve of specific wind turbine to calculate the annual mean wind speed, wind power density, utilization hour of specific wind turbine and annual energy output. Combined with daily extreme wind speed data from 1973 to 2011, the extreme wind with return period of 50-year was calculated using Peak over Threshold (POT) approach and Generalized Pareto distribution model. The analysis results show that the wind power density of Yongxing Island is less than expected, which is below $100\,W/m^2$ in a typical year. Moreover, the extreme wind with return period of 50-year is 42.2 m/s, which is significantly higher than reference wind speed (27 m/s) on the land.

Sun et al. (2017) investigated the wind energy and wind speed in the coastal waters of China based on the reanalysis data with horizontal resolution of $0.75° \times 0.75°$ from the European Center of Medium-range Weather Forecast (ECMWF) for the period 1979–2014. And Sun et al. (2017) used the Weather Research Forecast (WRF) model to simulate the influence of sea surface temperature (SST) increase and urbanization development on the impact of wind energy in east coast of China. Results show that: (1) In spring, the wind energy in Bohai gulf is obviously larger than that in other regions (coastal areas of East China, Southeastern China and the north of the South China Sea). In summer, the wind energy in Bohai gulf is clearly smaller than that in other regions and it is slightly larger in coastal areas of East China. In autumn, the wind energy is relatively abundant in coastal areas of southeastern China and the north of the South China Sea. In winter, the wind energy is similar in the four coastal regions of China. Besides, the wind speed presents an increasing trend in the coastal area of southeastern China in winter. However, it presents a slight decreasing trend in other areas throughout the year, but at a small scale. In terms of SST's influence on the wind speed, the trends may vary in different seasons and different areas: in spring, the wind speed may increase with SST rising in the Bohai Gulf, the Shandong Peninsula, the Beibu Gulf coast and the Hangzhou Bay. The same goes to most coastal areas in summer; the coastal areas of Southeastern China and East China in autumn and the coastal areas of Bohai Gulf and the north of the South China Sea in winter.

Waewasak et al. (2015) used simulated data and mapped out the wind energy at the height of 40, 80, 100 and 120 m in the Gulf of Thailand (high-spatial resolution of 200 m). And he also used the observation data of 28 meteorological towers to justify the validity of the simulated data. The result shows that in the Bay of Bangkok, the output of wind power reaches 6 TWh/yr with an amount of 3000 MW available under current technologies. In terms of the entire Gulf of Thailand, the annual total output is 15 TWh/yr with an available amount of 7000 MW.

Chang et al. (2014) used satellite data and model data to analyze the wind energy at the height of 10 m and 100 m around the surrounding ocean areas of the Hainan Island. The result shows that the wind power density is 400 to 600 W/m^2 at the height of 100 m in the east of Hainan Island.

Wan et al. (2018) used the ERA-Interim data to analyze the features of wind power in the South China Sea and found that wind power is relatively abundant at the Taiwan Strait, the Luzon Strait and the ocean areas southeastern of the Indo-China Peninsula.

Lip-Wah et al. (2012) used the satellite data to calculate the wind power in Malaysia and found that the wind speed is 6 to 7 m/s at a height of 50 m.

2.2 Researches on Wind Power of the Northern Indian Ocean

Nayyar and Zaigham (2013) calculated the wind power density of the southeastern coastal areas of Pakistan using wind data from United States National Aeronautics and Space Administration (NASA) and established a GIS-linked wind potential mode. The research found that there is abundant wind power in ocean areas near Karachi.

Murali et al. (2014) used the weekly wind resources from 1999 to 2009 to research the wind power along the coastal areas of India. The results show that at the targeted area, the wind speed is from 1.9 m/s to 10.2 m/s, and based on that, the research points out that the wind power is relatively abundant in Mumbai northeastern of the Arabian Sea.

Contestabile et al. (2017) analyzed the wind and wave energy of Maldives using data from ERA-Interim and found out that wave power density in this area was from 8.46 to 12.75 kW/m, and the wind power density is 80 W/m^2 and 160 W/m^2 at the height of 10 m and 100 m respectively.

Patel et al. (2019) found that the annual average wind power density reached 400 W/m^2 in the coastal areas of Tamil Nadu and the WPD 8 kW/m in the coastal areas of Gujarat.

Kumar et al. (2019) used the buoy observation to test the validity of the wind-mill resources of QuikSCAT, OSCAT, ASCAT-A and ASCAT-B, and found that the error of QuikSCAT and OSCAT is relatively small, being 0.15 m/s (2.4%) and 0.83 m/s (15.1%) respectively. Kumar also used linear regression to correct the satellite resources and carried out the researches on wind power. The results show that the wind power density of the west Arabian Sea is 450 to 550 W/m^2, which is significantly higher that of the Bay of Bengal.

Yip et al. (2016) used the MERRA resources and studied on the wind power over the Arabian Peninsula and found out that it is more dynamic than that over the Red Sea at the same latitude.

Yang et al. (2019) carried out joint research on the wind and the wave power over the North Pacific based on the ERA-Interim wind data and wave simulation data. The results showed that the water areas rich in wind power are located in the Somali waters and the Arabian sea.

Zheng et al. (2016) used the ERA-Interim wind data to comprehensively consider a series of key indicators such as wind power density, effective wind speed occurrence, energy level occurrences, energy direction, etc., and realized the wind energy climate characteristics of the Gwadar Port. The results show that the wind energy resources of the Gwadar Port are available all year round. The wind power density, effective wind speed occurrence, and energy level occurrences all show an obvious single peak characteristic in monthly changes. The peak value appears in April to May, and the trough appears in November to December. The annual average values are: wind power density of 121 W/m^2, effective wind speed occurrence of 43%, and energy level occurrence above 50 W/m^2 of 55%. It can be seen from the wind energy rose diagram that the wind energy of Gwadar Port is stable all the year round and contributes from the SW (southwest) wind. In February and November, the high wind power density above 400 W/m^2 and 500 W/m^2 was mainly contributed by the NNE (North–North-East) wind. In May and August, the high wind power density was mainly contributed by the SW (southwest) wind. The wind energy of the Gwadar Port has gone through two main stages. During 1979–2000, the wind power density fluctuated around 120 W/m^2, and the effective wind speed occurrence fluctuated around 45%; during 2001–2014, the wind power density was 110 W/m^2 fluctuates up and down, and the effective wind speed occurrence fluctuates around 40%. The frequency of strong winds above class 6 in Gwadar Port is within 1.5% all year round; the most frequent wind is class 3 (34.29%), followed by class 4 (28.32%) and level 2 (23.11%).

Zheng et al. (2017a, 2017b) further analyzed the historical change trends of a series of key indicators of wind energy in Gwadar Port, and realized the mid- and long-term forecast of wind energy in Gwadar Port, providing a technical approach for wind energy evaluation at key nodes of the Marine Silk Road. The results show that: (1) Gwadar Port has richer wind energy resources in summer than that in winter, and its stability in summer is significantly better than that that in winter. (2) During the past 36 years (1979–2014), the wind power density, effective wind speed occurrence, and energy level occurrence above 100 W/m^2 exhibit significant increasing trends, of -0.78 $W/(m^2 \cdot a)$, -0.21%/a, -0.22% %/a respectively. And the trend is mainly reflected in summer, while there is no significant change trend in winter. The stability of wind energy resources (coefficient of variation (Cv), monthly variability index (Mv), seasonal variability index (Sv)) has no significant change. (3) From the perspective of forecast values, the wind energy resources of the Gwadar Port remained at the same level as the multi-year average in 2015, and tended to be more abundant in 2016; from 2015 to 2016, the stability of wind energy resources is predicted to be slightly worse than the multi-year average status.

Zheng (2018a) took advance in analyzing the characteristics of wind energy of the twenty-first century Maritime Silk Road, by using the ERA-Interim wind data from the European center for Medium-Range Weather Forecasts. The value size of wind power density; available rate of wind energy; and richness, stability, and storage of wind energy were considered during the analysis. Results show that the twenty-first century Maritime Silk Road is rich in wind energy, especially the Somalia waters, the traditional especially the Somalia waters, the traditional gale center of the South China Sea and the Luzon Strait, followed by the Mannar Sea and the southeast waters of Sri Lanka. The above regions are advantageous in the size of wind power density, effective wind speed occurrence, energy level occurrence, energy storage, and coefficient of variation (Cv). However, these regions are disadvantageous in terms of monthly variability index (Mv) and seasonal variability index (Sv).

Zheng (2019) further calculated the long-term climatic trend of wind energy resources of the Maritime Silk Road, covering a series of key indicators such as wind power density, effective wind speed occurrence, energy level occurrence above 200 W/m^2, and resource stability, including the annual trend, the regional and seasonal differences of the long-term trend, and found that the wind energy resources of the Maritime Silk Road during the period 1979–2015 tended to be optimistic: the wind power density in most sea areas, the effective wind speed occurrence, and the energy level occurrence above 200 W/m^2 showed a significant increase or no significant trend, meaning that the wind energy trend to be more abundant or stable, only some scattered sea areas showed a significant decrease. The monthly variability index and seasonal variability index in most regions showed a significant decrease or no significant change trend, meaning that the energy stability trend to be better or stable, and only some scattered sea areas showed a significant increase.

2.3 Challenges and Resolutions

2.3.1 Detail Investigation on Climate Features of Wind Energy

The general exploration of wind power's climate features is fundamental to the energy classification, historical long term trend and short-term forecasting and mid-long-term projection (Zheng and Li 2018). And it further influences the siting of resource exploitation, business running and mid-long-term planning. In early stages, factors included in wind power evaluation were rather limited: including wind power density, storage and stability. However, in the actual exploration process, the resources' availability, richness and their sources are also of great importance. Zheng et al. (2012, 2013) and Zheng and Pan (2012) proposed two key indicators: energy level occurrences and effective wind speed occurrence to depict the richness, availability of resources. These two indicators are widely accepted by researchers domestic and aboard (Langodan et al. 2016). Stable wind power is good for power collection and transformation, and unstable one will not only impair efficiency of exploration and but also service life of the windmill. Thus, it's necessary to collect the joint 2019 occurrence of wind power density and wind directions to demonstrate the source of wind power.

Zheng and Li (2019) pointed out that the general exploration of wind power's climate features should systemically cover wind power density, availability, energy level occurrences, energy direction and its features (co-occurrence of wind power density and wind directions), stability (coefficient of variation (Cv), monthly variability index (Mv) and seasonal variability index (Sv)), storage (total storage, effective storage and technological storage) and demonstrate the spatial and temporal distribution of above key indicators. Here are the annual wind power density and availability (effective wind speed occurrence) as an example, as shown in Fig. 2.1.

2.3.2 Mapping of Macro–micro Scale Classification of Wind Power

Proper mapping of wind power classification is crucial to choose the best site for exploitation. In 2005, National Renewable Energy Laboratory (NREL 2005) in the United States mapped the classification of wind power globally based on the wind power density, which provided reference to the general location choice for its exploitation. However, regional differences of resource classification are not significant. For example, grade 7 covers vast sea areas under westerlies in the Northern and Southern Hemisphere and water areas along the Yangtzi River and the Peral River. Thus, such divisions couldn't guide us to choose which is the best place for wind power exploitation. Besides, current classification standards only include some factors of wind power; environmental risks and cost-effectiveness which have tremendous influence on siting are not taken into allowance. Thus, it is needed to establish a scheme which comprehensively consider the resource features, environmental risks, cost-effectiveness, and fully demonstrates the regional differences.

Using the Delphi method, Zheng et al. (2018a, 2018b) proposed a scheme comprehensively covering resource features, environmental risks and cost-effectiveness for the first time, which, to be more specific, includes 8 factors: wind/wave power density, availability, richness, coefficient of variation, monthly variability index, extreme wave height/extreme wave height, water depth and distance to coast. This scheme largely increased the efficiency of data collection, the safety of construction and the benefits of investments compared to traditional ones. Besides, this scheme illustrates

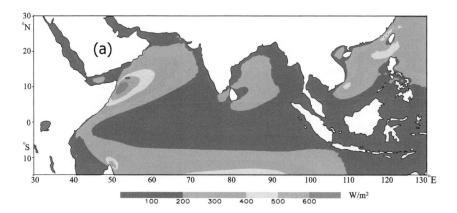

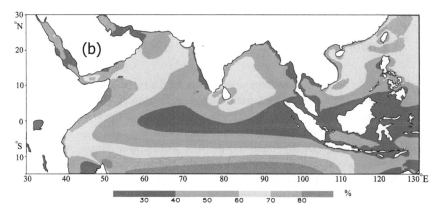

Fig. 2.1 Annual average wind power density **a** and annual effective wind speed occurrence **b** of the Maritime Silk Road (after Zheng and Li 2014)

the regional differences of resources of various grades, which is not just suitable for the mapping of resource classification over vast sea areas, but also for that over smaller areas, so as to provide scientific support for choosing the best location for wind power exploitation on a macro/micro scale. And the author and his group (Zheng et al. 2019) further proposed a new mapping concept, the dynamic self-adjustment classification for energy resources, which can be used support the building of remote islands and riffs, siting for commercial development and other purposes and guidance or technical support for classification mapping of wind energy, wave energy, and other new resources. Below is the mapping of wind power classification of the areas along the Maritime Silk Road. The wind energy classification of the Maritime Silk Road based on the new energy classification was presented in Fig. 2.2.

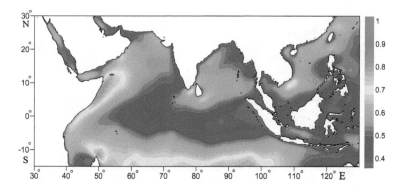

Fig. 2.2 Wind energy classification of the maritime silk road. A higher grade means a richer resource

2.3.3   Relationship Between Wind Power and Important Factors

Previous researches have done tremendous work on the relevance between meteorological/oceanic elements and important astrogeodetic factors. However, untill now, there are scarce researches on the relevance between wind power and important factors, which are crucial basis for the mid- and long-term wind power projections. Zheng et al. (2017a, 2017b) found out that the wind power density on the North Pacific is contemporaneously correlated with the North Atlantic Oscillation (NAO) with a positive–negative-positive relevance from high-altitude to low-latitude. The wind power density (lagging 3 months) is negatively relevant to NINO3 index. It is necessary to comprehensively analyze the relevance between key indicators of wind power (wind power density, availability, energy level occurrences, stability, etc.) and important factors (Arctic Oscillation (AO), Antarctic Oscillation (AAO), etc.) and its physical mechanism, so as to provide theoretic support to mid- and long-term projections. Hereby presents the relevance between the correlation index of the wind power density and the AAO index in the Maritime Silk Road, as shown in Fig. 2.3.

2.3.4   Short-Term Wind Power Forecasts

Short-term wind power forecasting guarantees the daily operation of wind turbines, increase the efficiency of wind power collection and its transformation, and provide precise guidance to the short-term electric power distribution. The commonly used forecasts models include Predictor forecasting system, eWind, West and Previento (Focken et al. 2001; Yu et al. 2006). Zheng et al. (2014) has interpreted and applied the wind field forecasts into the wind power forecasts of China seas and offered a new economical technology approach to wind power forecasts. There are three

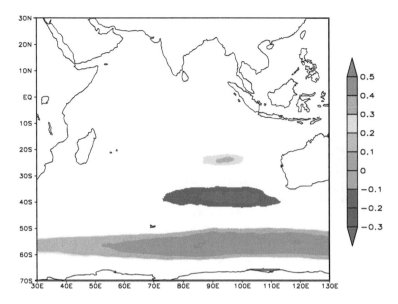

Fig. 2.3 Correlation between wind power density and AAO index of the maritime silk road. *Note* Only area significant at 95% level is colored

main short-term wind power forecasts: (1) Fully utilizing current climate forecasting products and combining it with wind power forecasts, so as to minimize the cost. (2) Combing mesoscale WRF or MM5 and CALMET and ARPS so as to adopt and WRF/CLAMET or WRF/ARPS modelling system. (3) Utilizing wind power forecasting apps domestic and aboard.

Besides, current short-term wind power forecast is mainly for wind power density and wind speed. Zheng et al. (2019a) pointed that short-term wave energy forecast should not only include the wave power density but also its availability and storage for the next few days. Furthermore, short-term wind power forecasts should include wind fields, wind power density, weekly availability, daily/weekly storage, resource direction and wind power forecasts at key points. Hereby presents the wind power density forecast map for water areas along the Maritime Silk Road, as shown in Fig. 2.4.

2.3.5 Long-Term Variation of Wind Power

Currently there are rich researches on the long-term variation of meteorological and oceanic elements. However, those on the long-tern variation of wind power are still scarce and mainly focus on wind power density and the changing of wind speed (Zheng et al. 2017a, 2017b). In the actual exploitation process, the energy stability is related to the efficiency of their collection and transformation and the service life of equipment, while the effective wind speed occurrence is linked to the energy

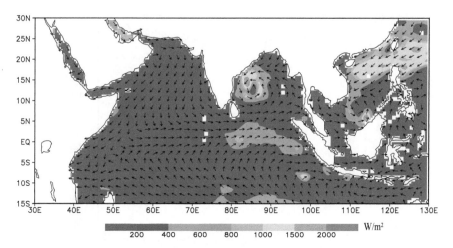

Fig. 2.4 Short-term forecast of wind power density (00:00 UTC, December 11, 2016) in the maritime silk road. *Note* Color represents the wind power density and arrow represents the wind direction

availability and the energy level occurrences to the energy richness. Thus, to analysis the long-term variation of wind power, it is necessary to fully calculate the long-term variation of a series of key factors such as wind power density, EWSO, ELO, Cv, Mv and Sv, to build a solid theoretic foundation for the mid- and long-term projection of wind power. Zheng et al. (2018b) used the ERA-Interim wind data, and calculated the trends of a series of factors from 1979 to 2015 along the Maritime Silk Road. Hereby presents the trends of the wind power density and EWSO, as shown in Figs. 2.5 and 2.6.

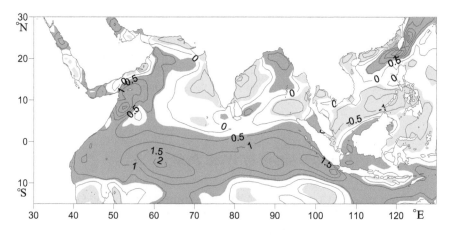

Fig. 2.5 Annual trend of wind power density along the maritime silk road. *Note* Green color area with increase trend significant at 95% level. Yellow color area with decrease trend significant at 95% level

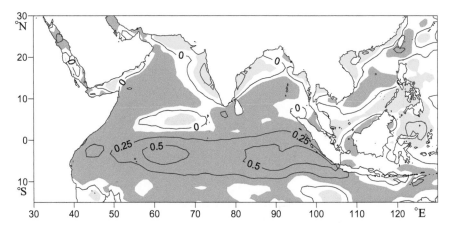

Fig. 2.6 Annual trend of effective wind speed occurrence along the maritime silk road. *Note* Green color area with increase trend significant at 95% level. Yellow color area with decrease trend significant at 95% level

2.3.6 Mid- and Long-Term Projection of Wind Power

When it comes to mid- and long-term planning of resource exploitation, the mid- and long-term projection is the mainstay. However, relevant researches are still scarce. Based on the artificial neural network, linear extension, Zheng et al. (2017a, 2017b) used ERA-Interim's data from 1979 to 2014 and conducted the mid- and long-term projection of wind power in Gwadar Port for 2015–2016. The mid- and long-term projection of the wind power density and the EWSO are shown Fig. 2.7. Zheng et al. (2019) also carried out long-term projection on the offshore wind power globally from 2080–2099 based on the Coupled Model Intercomparison Project (CMIP) data, including wind power density, EWSO and energy level occurrence over 200 W/m^2, then compared the result with the wind power from 1980 to 1999 and finally mapped the classification of the future wind power, focusing on the distribution and changing of areas with abundant resources. Above methods have provided technical approach to mid- and long-term projection.

Zheng et al. (2016) pointed out that in the future there are three ways of mid- and long-term projection: (1) Conducting statistical analysis on the relevance between the important astrogeodetic factors and the wind power, studying the physical mechanism so as to support the mid- and long-term projections on the wind power. (2) Carrying out mid- and long-term projections on the key index of the wind power based on the least square support vector machine, artificial neural network, Hilbert et al. (3) Using CMIP data to conduct mid- and long-term projections on a series of key factors of the wind power.

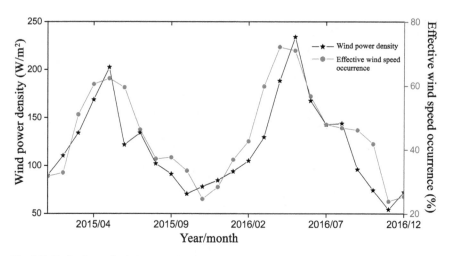

Fig. 2.7 Projections of wind power density and effective wind speed occurrence of the Gwadar Port for 2015–2016 based on the wind data for the period 1979–2014 (after Zheng et al. 2017a, 2017b)

2.3.7 Wind Power Evaluation on the Key Nodes

In most cases, remote islands are key supports for marine exploration. Building a series of stable and efficient key points are of great significance. As these islands are far away from the continent, electricity supply under such a condition has always been a global challenge. Diesel, delivered by ship, has served as a common solution, but it is not without its problems: the supply lines are too long and could easily be subjected to extreme weather; the ecological environment on such islands is too fragile to be restored once polluted by diesel engines. Thus, offshore wind energy and desalination programs could be an antidote to the power shortage and a blessing to island dwellers and countries along the Maritime Silk Road which are facing energy and environment crises.

Compared with traditional energy, offshore wind power eclipse in terms of ecological protection, availability and adaptive capacity. Exploiting the offshore wind power according to local conditions could help islands and reefs generate enough electric power while protecting the ecological environment. However, the wind power evaluation on key points are still scarce due to the shortage of resources, high demand of technology and lack of theoretical basis. Zheng et al. (2016) analyzed the climate features of the wind power at the Gwadar Port (as shown Figs. 2.8 and 2.9). In 2017, Zheng et al. (2017a, 2017b) analyzed the historical trends of a series of key indicators such as the wind power density, EWSO, ELO, gale occurrence and stability and further realize the mid- and long-term projections on the wind power of Gwadar Port.

In the future, we could refer to methods mentioned above and establish a widely adaptive assessment system of the islands and reefs wind power, including the climate

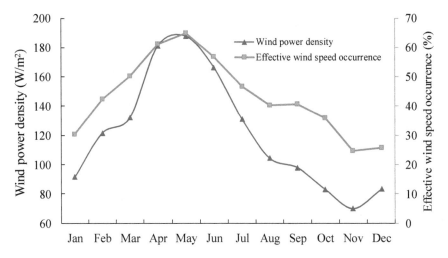

Fig. 2.8 Monthly variation of wind power density and effective wind speed occurrence of the Gwadar Port (after Zheng et al. 2016)

Fig. 2.9 Wind energy rose (co-occurrence of wind power density and wind direction) in the Gwadar Port (after Zheng et al. 2016)

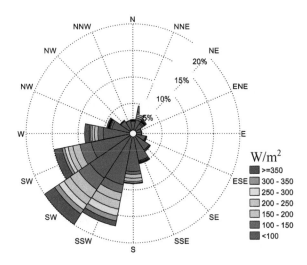

features of resources, the correlations between wind power and the important astro-geodetic factors, short-term forecasts, climatic long-term variation and long term projection of resources. Referring author's micro-scale mapping of classification (Zheng et al. 2019) and carry out dynamic and adaptive classification of wind power, so as to guarantee the accurate siting of their wind power exploitation, the running of business and mid- and long-term planning.

2.3.8   Construction of Wave Energy Resource Dataset

The production and sharing of marine data has become an important manifestation of comprehensive national strength. With the rapid development of observation technology, computer technology and calculation ability, the observation data, simulation data and reanalysis data are rapidly increasing. However, how to extracted the useful information about wind energy development from the original big data with large volume and low information density, and then building a wind energy dataset has become the key support to the industrialization and efficient deployment of wind energy. It is also a common challenge for global colleague. Now, the marine and meteorological raw data are relative abundant. However, the wind energy dataset is extremely scarce. The first temporal-spatial characteristics dataset of offshore wind energy resource for the 21st Century Maritime Silk Road was established by Zheng (2020), systematically including the wind power density (WPD), effective wind speed occurrence (EWSO), energy level occurrences, coefficient of variation (Cv), monthly variability index (Mv), seasonal variability index (Sv), total storage, effective storage and technological storage, which provide an important reference for the production of wind energy dataset.

References

Albani A, Ibrahim MZ, Yong KH (2018) Influence of the ENSO and monsoonal season on long-term wind energy potential in Malaysia. Energies 11:2965. https://doi.org/10.3390/en11112965

Carvalho D, Rocha A, Gómez-Gesteira M, Santos CS (2014) Offshore wind energy resource simulation forced by different reanalyses: comparison with observed data in the Iberian Peninsula. Appl Energy 134:57–64

Chadee XT, Clarke RM (2014) Large-scale wind energy potential of the Caribbean region using near-surface reanalysis data. Renew Sustain Energy Rev 30:45–58

Chang R, Zhu R, Badger M, Hasager CB, Xing XH, Jiang YR (2014) Offshore wind resources assessment from multiple satellite data and WRF modeling over South China Sea. Remote Sens 6:1–21

Chen XP, Wang KM, Zhang ZH, Zeng YD, Zhang Y (2017) An assessment of wind and wave climate as potential sources of renewable energy in the nearshore Shenzhen coastal zone of the South China Sea. Energy 134:789–801

Chen L, Xu W, Liu GY (2014) Wind resource assessment and extreme wind of 50-year recurrence in Xisha. Electric Power Construction 35(7):131–135

Contestabile P, Lauro ED, Galli P, Corselli C, Vicinanz D (2017) Offshore wind and wave energy assessment around malè and Magoodhoo Island (Maldives). Sustainability 9:613. https://doi.org/10.3390/su9040613

Dou RT, Xie Q (2020) The development scale and trend of China's offshore wind power based on grey forecasting model. Ocean Develop Manag 37(10):62–68

Focken U, Lange M, Waldl HP (2001) Previento—a wind power prediction system with an innovative upscaling algorithm. In: Proceedings of the European wind energy conference, copenhagen, Denmark, June 2–6, pp 826–829

Jiang J, Liu YX, Li MC (2014) Wind energy resources and wind power generation on islands and reefs in the South China Sea based on QuikSCAT wind data and landsat ETM+ Images. Resour Sci 36(1):139–147

Jiang Y, Wu MQ, Huang CJ, Niu Z (2000–2019) Collection of data on overseas power projects in the belt and road initiative (2000–2019). China Sci Data. https://doi.org/10.11922/csdata.2019.0069.zh

Kumar SVVA, Nagababu G, Kumar R (2019) Comparative study of offshore winds and wind energy production derived from multiple scatterometers and met buoys. Energy 185:599–611

Langodan S, Viswanadhapalli Y, Dasari HP, Knio O, Hoteit I (2016) A high-resolution assessment of wind and wave energy potentials in the red sea. Appl Energy 181:244–255

Le CH, Zhang J, Ding HY, Zhang PY, Wang GL (2020) Preliminary design of a submerged support structure for floating wind turbines. J Ocean Univer China 19(6):1265–1282

Lip-Wah H, Ibrahim S, Kasmin S, Omar CMC, Abdullah AM (2012) Review of offshore wind energy assessment and siting methodologies for offshore wind energy planning in Malaysia. Am Int J Contemp Res 2(12):72–85

Liu YC, Chen DY, Li SW, Chan PW (2018) Discerning the spatial variations in offshore wind resources along the coast of China via dynamic downscaling. Energy 160:582–596

Murali RM, Vidya PJ, Modi P, Kumar SJ (2014) Site selection for offshore wind farms along the Indian coast. Indian J Geo-Marine Sciences 43(7):1401–1406

National Renewable Energy Laboratory (NREL) QuikSCAT Annual wind power density at 10m. From(http://en.openei.org/w/index.php?title=File:QuikSCAT-_Annual_Wind_Power_Density_at_10m.pdf&page=1

Nayyar ZA, Zaigham NA (2013) Assessment of wind potential in Southeastern part of Pakistan along coastal belt of Arabian Sea. Arab J Sci Eng 38:1917–1927

Niu F, Chen S, Li Y, Ha HC (2019) Prospects for the development of British offshore wind power industry: analysis based on the perspective of CfD. Ocean Develop Manag 36(9):68–74

Patel RP, Nagababu G, Jani HK, Kachhwaha SS (2019) Wind and wave energy resource assessment along shallow water region of indian coast. In: Twelve international conference on thermal engineering: theory and applications, February 23–26 Gandhinagar, India

Sun YT, Nian XY, Min JZ (2017) Distribution characteristics of wind energy along the coast of China and numerical simulation on impact factors. Trans Atmos Sci 40(6):823–832

Waewsak J, Landry M, Gagnon Y (2015) Offshore wind power potential of the Gulf of Thailand. Renew Energy 81:609–626

Wan Y, Fan CQ, Dai YS, Li LG, Sun Wei F, Zhou P, Qu XJ (2018) Assessment of the joint development potential of wave and wind energy in the South China Sea. Energies 11:398. https://doi.org/10.3390/en11020398

Yang SB, Duan SH, Fan LL, Zheng CW, Li XF, Li HY, Xu JJ, Wang Q, Feng M (2019) 10-year wind and wave energy assessment in the North Indian Ocean. Energies 12:3835. https://doi.org/10.3390/en12203835

Yip CMA, Gunturu UB, Stenchikov GL (2016) Wind resource characterization in the Arabian Peninsula. Appl Energy 164:826–836

Youm I, Sarr J, Sall M, Ndiaye N, Kane MM (2005) Analysis of wind data and wind energy potential along the northern coast of Senegal. Renew Energy 8:95–108

Yu W, Benoit R, Girard C (2006) Wind Energy Simulation Toolkit (WEST): a wind mapping system for use by the wind-energy industry. Wind Eng 30(1):15–33

Zheng CW (2011) Wave energy and other renewable energy resources in South China Sea: advantages and disadvantages. J Subtropical Res Environ 6(3):76–81

Zheng C (2018a) Wind energy evaluation of the 21st century maritime silk road. J Harbin Eng Univer 39(1):16–22.

Zheng CW (2018b) 21st Century Maritime Silk Road: wave energy evaluation and decision and proposal of the Sri Lankan waters. J Harbin Eng Univer 39(4):614–621

Zheng CW (2018c) Energy predicament and countermeasure of key junctions of the 21st century maritime silk road. Pacific J 26(7):71–78

Zheng CW (2018d) Wind energy trend in the 21st century maritime silk road. J Harbin Eng Univer 39(3):399–405

Zheng CW, Gao Y, Chen X (2017a) Climatic long term trend and prediction of the wind energy resource in the Gwadar Port. Acta Scientiarum Naturalium Universitatis Pekinensis 53(4):617–626

Zheng CW, Gao CZ, Gao Y (2019) Climate feature and long term trend analysis of wave energy resource of 21st Century Maritime Silk Road. Acta Energiae Solaris Sinica 40(6):1487–1493

Zheng CW, Li CY (2015) Development of the islands and reefs in the South China Sea: wind power and wave power generation. Periodical of Ocean Univer China 45(9):7–14

Zheng CW, Li CY (2015) Development of the islands and reefs in the South China Sea: wind climate and wave climate analysis. Periodical of Ocean University of China 45(9):1–6

Zheng CW, Li CY (2017) 21st century maritime silk road: big data construction of new marine resources: wave energy as a case study. Ocean Develop Manage 34(12):61–65

Zheng CW, Li CY (2018) An overview and suggestions on the difficulty of site selection for marine new energy power plant—wave energy as a case study. J Harbin Eng Univer 39(2):200–206

Zheng CW, Li CY (2019) Evaluation of new marine energy for the maritime silk road from the perspective of maritime power. J Harbin Eng Univer 41(2):175–183

Zheng CW, Li XY, Luo X, Chen X, Qian YH, Zhang ZH, Gao ZS, Du ZB, Gao YB, Chen YG (2019c) Projection of future global offshore wind energy resources using CMIP data. Atmos Ocean 57(2):134–148

Zheng CW, Li CY, Pan J, Liu MY, Xia LL (2016a) An overview of global ocean wind energy resource evaluations. Renew Sustain Energy Rev 53:1240–1251

Zheng CW, Li CY, Xu JJ (2019d) Micro-scale classification of offshore wind energy resource—a case study of the New Zealand. J Clean Prod 226:133–141

Zheng CW, Li CY, Yang Y, Chen X (2016) Analysis of wind energy resource in the Pakistan's Gwadar port. J Xiamen Univer (natural Science Edition) 55(2):210–215

Zheng CW, Pan J (2012) Wind energy resources assessment in global ocean. J Nat Resour 27(3):364–371

Zheng CW, Pan J, Li JX (2013) Assessing the China Sea wind energy and wave energy resources from 1988 to 2009. Ocean Eng 65:39–48

Zheng CW, Xiao ZN, Peng YH, Li CY, Du ZB (2018c) Rezoning global offshore wind energy resources. Renew Energy 129:1–11

Zheng CW, Zhou L, Song S (2014) Forecasting of the China Sea wind energy density. J Guangdong Ocean Univer 31(1):71–77

Zheng CW, Zhuang H, Li X, Li XQ (2012) Wind energy and wave energy resources assessment in the East China Sea and South China Sea. Sci China Technol Sci 55(1):163–173

Zheng CW, Li CY, Wu HL, Wang M (2018a) In: 21st century maritime silk road: construction of remote islands and reefs. Springer

Zheng CW, Xiao ZN, Zhou W, Chen XB, Chen X (2018b) In: 21st century maritime silk road: a peaceful way forward. Springer

Zheng CW, Xu JJ, Zhao C, Wang Q. (2019a) In: 21st century maritime silk road: wave energy resource evaluation. Springer

Zheng CW, Li CY, Li X (2017b) Recent decadal trend in the North Atlantic wind energy resources. Adv Meteorol 7257492:8. https://doi.org/10.1155/2017/7257492

Zheng CW, Chen X, Sun W (2018) Accelerate the development of a disciplinary system on the Maritime Silk Road. China Ocean News 2967(A2)

Zheng CW (2020) Temporal-spatial characteristics dataset of offshore wind energy resource for the 21st Century Maritime Silk Road. China Sci Data 5(4):(2020–12–28). https://doi.org/10.11922/csdata.2020.0097.zh

Zhou C, Huang YJ (2016) Assessment of wave energy resources in an offshore wind farm of Guangdong. Southern Energy Construct 3(4):119–122

Chapter 3
Climatic Temporal-Spatial Distribution of Offshore Wind Energy in the Maritime Silk Road

Previous researchers have done tremendous work studying the climate features of wind energy across global oceans. However, the studies on the climate features along the Maritime Silk Road are still scarce overall. Besides, the early-stage researches focused only on elements such as wind power density, storage and stability, without considering other key indicators such as the availability, richness and energy direction.

In 2015, Zheng and Li (2015a; b) analyzed the wind climate and the wind energy features of islands and reefs in the South China Sea. As the South China Sea is one of the key areas along the Maritime Silk Road, this analysis could facilitate its implementation. In 2016, based on the ERA-Interim wind field, Zheng et al. (2016a) groundbreakingly estimated the climate features of the wind power at the Gwadar Port in Pakistan, comprehensively considering a series of key indicators such as wind power density, EWSO, energy level occurrences and the energy direction and provided a technical approach to the wind power assessment for the key nodes along the Maritime Silk Road. In 2016, Zheng et al. (2016b) summarized the features and development conditions of resources in countries along the Maritime Silk Road, focusing on the wave power and the offshore wind power. In 2017, Zheng et al. (2017a) further analyzed the historical climatic trend of a series of key indicators of wind energy in the Gwadar Port, and carried out a mid- and long-term projection of wind energy, which provided a scientific support to the wind energy projection. In 2017, Zheng and Li (2017a) pointed out that while the data on meteorology and ocean are abundant, there are still few on marine energy which are urgently needed for resource assessment and engineering design. Zheng and Li (2017a) carried out a case study based on the wave energy dataset and pointed out that data should cover 6 modules: climatic feature of wave energy, energy classification, short-term forecast of the resource, features of swell energy, long-term trend and mid- and long-term projection of resources. And based on the 3D network and the temporal order, they also digitalized and stored the wave energy data, ensured its quality and 4D-visualized it. Based on the dataset, it is necessary to build a comprehensive application platform which is easy to use, complete in theoretic system and caters to actual needs (Zheng et al. 2017b), and could provide intellectual support to researchers and engineers. In 2018, Zheng (2018a) expressed the advantages of marine new energy in the construction of remotes islands and reefs. He also talked out the difficulties faced by current studies on the new energy in detail, including the collection of original data, the establishment of new energy assessment system for islands and reefs, establishment of short-term forecasts and mid- and long-term projection models, dataset construction of marine new energy, and provided countermeasure, thus to effectively guarantee site selection, daily running and mid- and long-term planning of the wave power generation, offshore wind power generation and sea water desalination. In 2018, based on resources from the ERA-Interim wind field, Zheng (2018b) detailed analyzed the climate features of the wind power along the Maritime Silk Road, comprehensively considering a series of key parameters, including the wind power density, availability, richness, stability and storage.

Chong-wei Zheng's team carried out the offshore wind power assessment at key nodes and vital areas, and then extended the assessment to the entire water areas along the Maritime Silk Road, which provided theoretic support to the offshore wind power exploitation and foundation for the dataset on offshore wind power resources. On the basis of our previous studies, this chapter used the ERA-Interim wind data from the European Centre for Medium-Range Weather Forecasts (ECMWF) and carried out a detailed research on the climate features of the wind power along the Maritime Silk Road, considering a whole package of key indicators such as wind power density, EWSO, energy level occurrences, resource stability and energy storage, to make contribution to easing the energy and environmental crisis, offer technical support to the construction of remote islands and reefs and promote the development the Maritime Silk Road.

3.1   Data and Methods

3.1.1   Data

The commonly used sea surface wind fields include the 40-yr ECMWF Re-Analysis (ERA-40), National Centers for Enviromental Prediction (NCEP), QuikSCAT/NCEP mixed field, CCMP and ERA-Interim (Table 3.1), while the last one excels in terms of spatial and temporal resolution and temporal order. In this chapter, we would further analyze the wind power along the Maritime Silk Road in detail based on ERA-Interim wind data. As a new product released after ERA-40, ERA-Interim adopts a meteorological model with higher spatial resolution and a much better data application and assimilation system. The ERA-Interim also adopts the advanced 12-hourly 4d Var with data including satellite resources, regular observation resources and model resources. It is not only a precursor to a revised extended reanalysis product to replace ERA-40, but also makes use of much archived data not available to the ERA-40. Covering the period from January 1st, 1979, its spatial range covers

Table 3.1 Commonly used wind data

Wind data	Resolution		Time series	Temporal space
	spatial	Temporal		
ERA-40	2.5° × 2.5°	6 h	1957.09 – 2002.08	87.5°S–87.5°N, 0°–57.5°E
NCEP	1° × 1°	6 h	1999.07 –	90°S–0°N, 0°–59°E
QN	0.5° × 0.5°	6 h	1999.08–2009.07	88°S–8°N, 0°E–60°E
CCMP	0.25° × 0.25°	6 h	1987.07–	78.375°S–78.375°N, 0.125°E–359.875°E
ERA-Interim	0.125° × 0.125°, …, 2.5° × 2.5°	6 h	1979.01 –	90°S– 90°N, 180°W–80°E

from 90°S to 90°N, 180°W to 180°E with spatial resolution ranging from 0.125° × 0.125°, 0.25° × 0.25°, 0.5° × 0.5°, 0.75° × 0.75°, ..., 2.5° × 2.5°. We chose the ERA-Interim from 1979 to 2015 with a spatial resolution of 0.25° × 0.25°. Overall, the ERA-Interim has relatively higher accuracy and is widely applied (Dee et al. 2011; Bao and Zhang 2013; Zheng 2020).

3.1.2 Method

This chapter mainly carries out a detailed research on the climatic temporal-spatial distribution of the wind power along the Maritime Silk Road. Based on the 6-hourly ERA-Interim data from 1979 to 2015, we could calculate the 6-hourly wind power density over a period of 37 years along the Road. We carried out a statistical analysis on the climate features of the wind power along the Maritime Silk Road according to the above data on the wind power density and wind speed. Energy availability and energy level occurrences of wind energy are defined. And we also systematically covered a whole package of key indicators' temporal and spatial distribution such as wind power density, availability, energy level occurrences, Mv, Sv and storage (total storage, exploitable storage and technological storage). The equation is as follows:

(1) The definition of wind power density and its calculation: Wind power density is defined as the mean annual power available per square meter of swept area of a turbine, with calculation as follows (Li et al. 2008; Zheng et al. 2012),

$$W = \frac{1}{2}\rho V^3 \qquad (3.1)$$

In this equation, W is the wind power density (W/m^2), V is the wind speed (m/s), ρ is the surface air density (kg/m^3) while the last one is usually 1.225 kg/m^3

under standard conditions for temperature and pressure with an altitude below 500 m (Capps and Zender 2010; Zheng et al. 2019a).

(2) Definition of wind power availability and its equation: Since 2011, Zheng et al. (2011) has defined the availability of wave energy and continued to improve this parameter. In 2012, Zheng and his team suggested that effective wind speed occurrence (EWSO) could be used to reflect the amount of wind power available (Zheng et al. 2012; Zheng et al. 2019b). Wind power availability is defined as the EWSO during the exploitation process. During the exploitation, it is recognized that the wind speed between 5 and 25 m/s is defined as the effective wind speed of wind energy development, the optimal speed for collection and transformation of wind power. The effective wind speed could also be defined as speed between 3 and 25 m/s under other criteria. Generally, offshore wind power is more abundant than that on land, and hence the former standard of wind speed between 5 and 25 m/s is adopted hereby. The equation of wind power availability is as follows:

$$EWSO = \frac{t_1}{T} \times 100\% \qquad (3.2)$$

In Eq. (3.2), EWSO is the effective wind speed occurrence (%), t_1 is the number of time steps with wind speed 5–25 m/s. T represents the total number of time steps analyzed.

(3) The definition of energy level occurrences: Since 2011, Zheng and Li (2011) groundbreakingly proposed the wave energy level occurrences, to depict the richness of wave power. In 2012, Zheng proposed the wind energy level occurrences to depict the wind power richness. In the early stage, this index mainly included the available level occurrence (ALO, occurrence of wind power density greater than 100 W/m^2) and rich level occurrence (RLO, occurrence of wind power density greater than 200 W/m^2). This chapter improves and further develops the wind energy level occurrences and set five level standards: ALO, moderate level occurrence (MLO, occurrence of WPD greater than 150 W/m^2), RLO, excellent level occurrence (ELO, occurrence of WPD greater than 300 W/m^2), superb level occurrence (SLO, occurrence of WPD greater than 400 W/m^2, as shown in Table 3.2.

The equation of energy level occurrences is as follows (Zheng et al. 2019b):

Table 3.2 Definition energy level occurrence of wind energy

Parameter	Abbreviation	Definition
Available level occurrence	ALO	Occurrence of WPD greater than 100 W/m^2
Moderate level occurrence	MLO	Occurrence of WPD greater than 150 W/m^2
Rich level occurrence	RLO	Occurrence of WPD greater than 200 W/m^2
Excellent level occurrence	ELO	Occurrence of WPD greater than 300 W/m^2
Superb level occurrence	SLO	Occurrence of WPD greater than 400 W/m^2

$$ALO = \frac{t_2}{T} \times 100\% \tag{3.3}$$

$$MLO = \frac{t_3}{T} \times 100\% \tag{3.4}$$

$$RLO = \frac{t_4}{T} \times 100\% \tag{3.5}$$

$$ELO = \frac{t_5}{T} \times 100\% \tag{3.6}$$

$$SLO = \frac{t_6}{T} \times 100\% \tag{3.7}$$

where, t_2, t_3, t_4, t_5, t_6 represent the time steps with wind power density greater than 100, 150, 200, 300, 400 W/m^2 separately. T represents the total number of time steps analyzed.

This chapter first analyzes the seasonal features of the wind power density, taking February, May, August and November as representative months for each season. Besides, the three key parameters that the author has proposed early-on, the effective wind speed occurrence, the occurrence level above 100 and 200 W/m^2, could effectively demonstrate the availability and the richness of the wind power. During the exploitation process, it is generally recognized that the wind speed between 5 and 25 m/s is the best for collection and transformation of the wind power (Miao et al. 2012; Zheng and Pan 2014). And the wind speed of that range is also defined as the effective wind speed. Generally, an area with wind power density over 200 W/m^2 is considered to be rich in that resources. Thus, using the 6-hourly ERA-Interim data and wind power density over a period of 37 years, the author counted the EWSO for a whole year, the energy level occurrences over 100 and 200 W/m^2, the stability and storage of wind power. The author calculated the Cv, the Mv and the Sv to exhibit the stability of wind power, conducted quantitative calculation on the amount in total, available and technically exploitable storage of each 0.25° × 0.25°grid point, and carried out a detailed study on the climate features of the wind power through a series of temporal and spatial analyses.

3.2 Wind Power Density

Wind power density is the clear demonstration of the richness of the wind power. Thus, we shall discuss about the temporal and spatial distribution features of wind power density first. Based on the 6-hourly ERA-Interim wind field data from 1979 to 2015 and the calculation method of wind power density, we calculated the data on the 6-hourly wind power density during this time period. Then we calculated the mean values of wind power density in February 1979 by averaging the wind power density from 00:00 February 1, 1979 to 18:00 February 28, 1979. Similarly, the mean values of wind power density in each February for the period 1979–2015 was obtained. Then the multi-year average value of wind power density in February was calculated. Using the same method, the multi-year average values of wind power density in May, August, November was calculated were obtained, as shown in Fig. 3.1.

The wind power density along the Maritime Silk Road has shown significant seasonal difference: In the South China Sea, February and November have the highest wind power density, followed by August, and May has the lowest. In the Bay of Bengal, August has the highest, followed by May, and February and November the lowest. In the Arabian Sea, August also eclipses all other 11 months in terms of wind power density, with May at the second place and February at the bottom. To put it more specifically:

In February (represents winter, same as below), the cold airs have momentous impact on the South China Sea and because of that, the wind power density is above 200 W/m^2 in most of the South China Sea. The strong winds with the wind power density up to 500–600 W/m^2 are mainly found among the Luzon Strait and the southeastern sea area of Indo-China Peninsula running in a northwest-southeast direction. Overall, the wind power density of the Beibu Gulf is above 100 W/m^2 with that at its center over 200 W/m^2. The Bay of Bengal is relatively lacking in wind power during winter with density basically within 100 W/m^2 at its north. The cold airs have less impact on the north Indian Ocean and makes its wind power density much less than that of the South China Sea. During winter, the wind power density of the Arabian Sea is mostly less than 200 W/m^2, while for the Bay of Bengal and the tropical Indian Ocean mostly below 100 W/m^2 with the relatively strong wind located at the coastal areas of eastern Somali (around 300 W/m^2). Besides, the wind power density of eastern Sri Lanka is from 100 to 200 W/m^2.

In May (representing Spring, same as below), which is in the transition phase of monsoons, overall, the Bay of Bengal has the highest wind power density, followed by the Arabian Sea and the South China Sea has the smallest. During spring, the southwest monsoon prevails in the North Indian Ocean with the cold air fading away. For most areas in the Arabian Sea, the wind power density is over 100 W/m^2, with the contour line running in a northeast-southwest direction. The maximum zone is found in the eastern sea areas of the Somali, most of the Bay of Bengal and the southeastern sea areas of Sri Lanka with the wind power density reaching 500, 150 and 450 W/m^2 respectively. In the northeastern coastal areas of Sri Lanka, there is a small area which has relatively low wind power density as the landscape of the island has shielded the southwestern monsoon. In spring, the wind power density in the South China Sea is relatively low overall because the cold air of wind has already gone while the southwest monsoon is still on its way. The wind power density of vast sea areas in the middle and the southern parts of the South China Sea is within 100 W/m^2 and only that of the northern part reaches 100–150 W/m^2. Interestingly, the wind power density in the south Indian Ocean increased significantly compared with that back in February, which is above 200 W/m^2.

In August (representing summer), with the southwest monsoon in dominance, the wind power density of the Arabian sea is basically over 400 W/m^2 (1600 W/m^2 at the center which is around the coastal areas of Somali) while that of the Bay of Bengal is

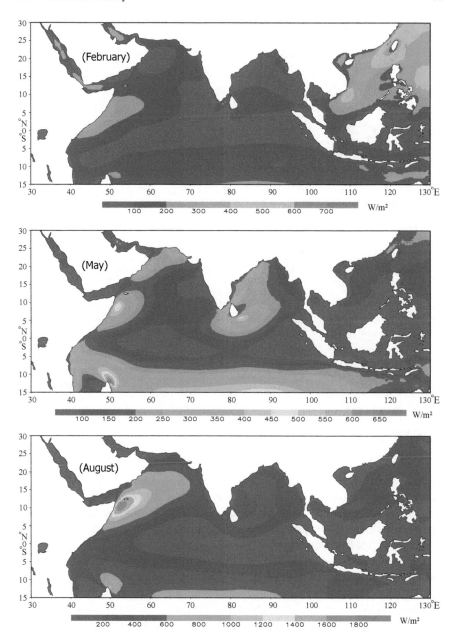

Fig. 3.1 Multi-year average wind power density in February, May, August, November and annual mean value along the maritime silk road (after Zheng 2018b)

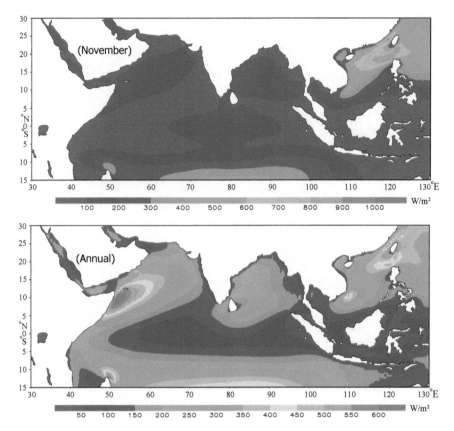

Fig. 3.1 (continued)

over 200 W/m^2. The value of the South China Sea is obviously lower than that of the North Pacific, and only that of the middle area is over 200 W/m^2. The strong winds are distributed around the South China Sea. And from the contour line, the Arabian Sea has the strongest southeastern monsoon, followed by the Bay of Bengal and the South China Sea has the weakest.

In November (representing autumn, same as below), the southwest monsoon in the north Indian Ocean has basically given way to the cold air which is still quite weak though. As a result, the wind power density for most areas in the north Indian Ocean is below 200 W/m^2. In the South China Sea, the cold air is obviously stronger than that in the north Indian Ocean with the wind power density basically over 200 W/m^2 in the middle and northern parts and the strong winds, with the wind power density up to 800 W/m^2, can be found in the north, especially the Luzon Strait and its western sea areas.

The wind power density along the Maritime Silk Road has shown clear regional differences from the annual average: the Arabian Sea has the highest wind power density accounting for 200 W/m^2 with the contour line running in a northeast-southwest direction; the South China Sea has it between 150–450 W/m^2 with the contour line running in a northeast-southwest direction; the wind power density in the Bay of Bengal is less than that in the Arabian Sea and the South China Sea but still above 150 W/m^2. The strong winds can be found in three places: the sea areas around Somali (wind power density over 400, 600 W/m^2 and above at the center), the southeastern sea areas of the Indo-China Peninsula (over 350 W/m^2) and the Luzon Strait (over 350 W/m^2). It is noteworthy that the Manarra Sea, which is between Sri Lanka and the Indian Peninsula and the southeastern sea areas of Sri Lanka, is rich in wind power with the density exceeding 250 W/m^2. During wind power exploitation, areas with the wind power density over 200 W/m^2 are considered to be rich in this resource, a criterion which fits the three areas mentioned above. Thus, these areas are crucial to the building of the Maritime Silk Road.

3.3   Availability of Wind Energy

The EWSO is a good indicator for the availability of the wind power resources. Hereby, we used the 6-hourly ERA-Interim wind field data from 1979 to 2015 and analyzed the EWSO along the Road in each representative month and for the whole year. For example, we used the 6-hourly wind field data of February, 1979 and calculated the EWHO on each 0.25° × 0.25°grid point, and then we can calculate the EWHO of each February from 1979 to 2015. Then the multi-year average EWHO in February was obtained. And the same could be applied to May, August, November and the whole year (as shown in Fig. 3.2).

The availability of the wind power along the Maritime Silk Road has shown great seasonal differences: in the South China Sea, February and November have the highest availability, followed by August and May has the smallest. In the Bay of Bengal, August has the highest, followed by May and November, and February comes at the end. In the Arabian Sea, August has the highest, followed by February and November, and May has the smallest. To put it more specifically:

In February, the EWSO in the South China Sea is generally high: most areas have it above 70%, which shows that this region is rich in availability of wind power during this season. For over half of the Arabian Sea, the EWSO is over 40%. The strong winds are found in the water areas around Somali with an EWSO above 95% at the center. Compared with these two regions, the EWSO of the Bay of Bengal is clearly lower, which is less than that of the inlet (30%), the outlet (60%) and the Manarra sea (which is between Sri Lanka and the Indian Peninsula).

In May, the EWSO is declining from the west (the Arabic Sea) to the east (the South China Sea). No matter it is in the Arabian Sea, the Bay of Bengal or the South China Sea, the EWSO of the northwestern sea areas is obviously higher than that of the southeastern. Moreover, the EWSO in the tropical south Indian Ocean increased drastically, accounting for 90% at the center.

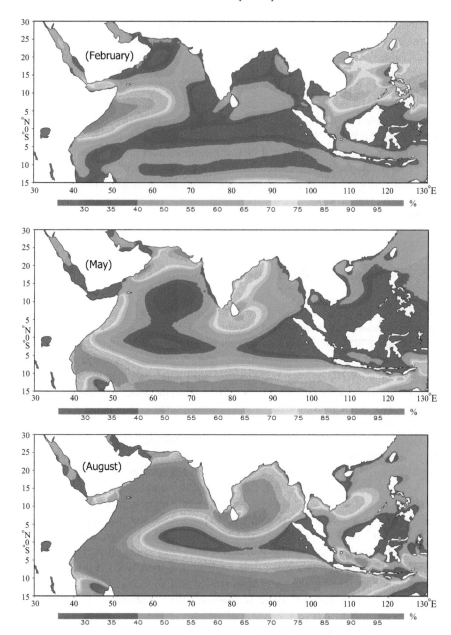

Fig. 3.2 Multi-year average effective wind speed occurrence in February, May, August, November and annual mean value along the maritime silk road (after Zheng 2018b)

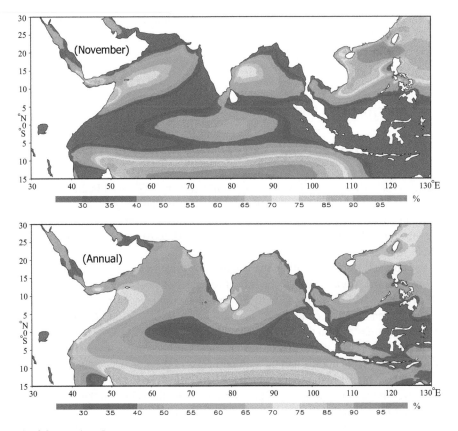

Fig. 3.2 (continued)

In August, influenced by the strong southwest monsoon, the EWSO of the water areas around Somali and the Arabian Sea is basically above 95%, while that of the Bay of Bengal is generally over 70%. Nevertheless, the monsoon in the South China Sea is weaker than that in the Arabian Sea and the Bay of Bengal. Thus, comparatively the EWSO of this water area is lower but still over 50%. And the EWSO of the tropical south Indian Ocean keeps growing and is generally over 95%.

In November, for areas in the north of 10°N in the South China Sea, EWSO is basically above 70%, where the highest occurrences, which are over 90%, can be found in the Luzon Strait, while the values of the Arabian Sea and that of the Bay of Bengal are both comparatively smaller than that in the South China Sea. The tropical south Indian Ocean still has relatively occurrences.

From the annual mean EWSO, the availability of wind energy along the Maritime Silk Road has also shown significant regional differences. The Arabian Sea, the Bay of Bengal and the South China Sea have the EWSO over 50% showing that the availability of wind power is promising overall. There are three obvious maximum centers: the sea areas around Somali (70% and above, 85% and above at its center), the southeastern sea areas around the Indo-China Peninsula (over 70%), the Luzon Strait (over 70%). And the Manarr sea and the southeastern sea areas of Sri Lanka are relative large center (over 65%).

3.4 Energy Level Occurrences

3.4.1 Occurrences Wind Power Density Above 100 W/m²

Based on the 6-hourly wind power density in February 1979, we obtained the ALO on each $0.25° \times 0.25°$grid point, and under the same method, we calculated the ALO of each February from 1979 to 2015. Then the multi-year average ALO in February was obtained. And the same could be applied to May, August, November and the whole year, as shown in Fig. 3.3.

In February, the South China Sea is under the greatest impact from the cold air, followed by the Arabian Sea and the Bay of Bengal. For most water areas in the South China Sea, the ALO is over 60% with the zone of maximum occurrences (over 70 and 80% at its center) located at the vast sea areas of southeastern Indo-China Peninsula and the sea areas around the Taiwan island (over 70%). The ALO of the Beibu Gulf is basically over 50% while that of the Thailand Gulf is comparatively lower which is within 20% in the middle and the northern parts of the gulf. The values of the Bay of Bengal are higher in the south, which are generally over 30 and 50% at the center, and lower in the north, which are less than 20%. And the values of the Arabian Sea go higher in the southwest (over 50%) and lower in the northeast. The ALO of the sea areas around Somali is above 90%, but the for most areas in the northeast of the Arabian Sea, it is less than 40%.

In May, overall, the Bay of Bengal has the highest ALO, followed by the Arabian Sea and the South China Sea is lagging behind. In the Bay of Bengal, most areas are over 40% with a contour line higher in the west and lower in the east. In the coastal areas of the Indian Peninsula, the contour line is basically in line with the coast with the ALO above 60%. The southeastern Sri Lanka has high occurrences which are over 70 and 80% at the center. In the Arabic Sea, occurrences are within 40% in the middle and over 70% at the top and above 60% around Somali. In the South China Sea, for most areas, the ALO is less than 50% and the maximum zones are distributed at the northern part of the Sea, the Beibu Gulf and the Thailand Gulf. In the South Indian Ocean, the ALO of water areas at low latitudes is basically over 50 and 90% at the center.

In August, the Arabian Sea has the highest ALO, followed by the Bay of Bengal, and the South China Sea comes at the end. Under the strong southwest monsoon, the north Indian Ocean reached its summit for ALO. In the Arabian Sea, the number is basically over 90% with the contour line going in a southwest-northeast direction, and in the Bay of Bengal, for most sea areas it is over 60 and up to 90% at the center. There is a small minimum zone at the northeastern coastal areas of Sri Lanka as the

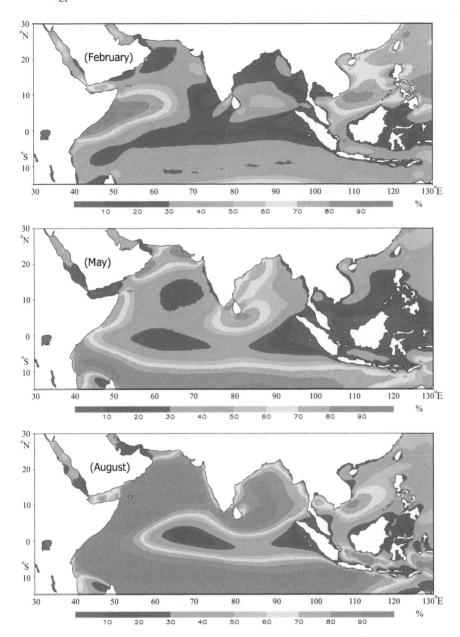

Fig. 3.3 Multi-year average occurrence of wind power density above 100 W/m^2 in February, May, August, November and annual mean value along the maritime silk road

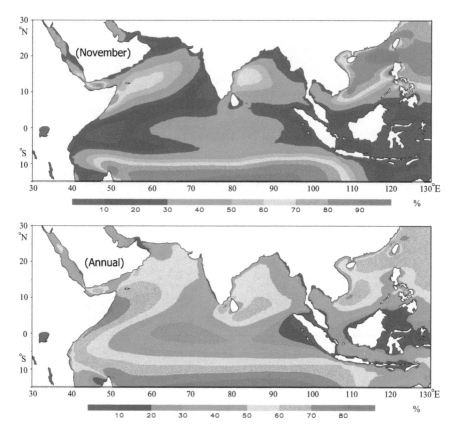

Fig. 3.3 (continued)

monsoon is blocked due to the landscape. In the South China Sea, the maximum center is located at the southeastern sea areas around the Indo-China Peninsula. Besides, the Thailand Gulf has relatively higher occurrences, generally over 60% and 70% at the center while the figure for the northern part of the South China Sea is around 40%.

In November, from the South China Sea to the North Indian Ocean, the southwestern monsoon has already stepped down and gave the floor to the cold airs. During this season, the cold airs in the north Indian Ocean are clearly weaker than that in the South China Sea, which also is reflected upon the ALO of these two regions. The ALO contour line runs in a northeast-southwest direction, with the maximum center distributed at the Luzon Strait and its western sea areas (over 90%). As for the middle and the northern parts of the South China Sea, the ALO, which is over 50%, is obviously higher than that for the southern part. The Beibu Gulf has it over 60%, while the Thailand Gulf less than 40% due to the weak cold air.

For most sea areas along the Maritime Silk Road, the annual average ALO is over 50%, which is a positive signal. In the Arabian Sea, the contour line runs in a northeast-southwest direction and the numbers decline in an opposite direction. The maximum center is located at the sea areas around Somali (over 70%). In the Bay of Bengal, there are two areas that have relatively high occurrences, which are both over 60%: the Manaar Gulf and the east of Sri Lanka, the latter which is in a band shape going in a southwest-northeast direction. In the South China Sea, the ALO for the middle and the northern parts is over 50%. The maximum zone is distributed at the north South China Sea and the traditional gale center of the South China Sea with values over 60 and 70% at the center. The ALO is relatively small in the Thailand Gulf and the middle and the southern parts of the South China Sea with values less than 40%.

From seasonal differences, in the Arabian Sea, August has the highest ALO, followed by February, and May comes at last, while in the Bay of Bengal it is August takes the first position and May that takes the second position and February is lagging behind. In the sea areas of Sri Lanka, August has the largest ALO, followed by the May and November comes at the end, while in the South China Sea November has the highest ALO, followed by February and May has the smallest.

3.4.2 Occurrences Wind Power Density Above 150 W/m^2

Based on the 6-hourly wind power density in February in 1979, we counted the MLO on each $0.25° \times 0.25°$bin in this month. And under the same method, we calculated the MLO in each February from 1979 to 2015. Then the multi-year average MLO in February was obtained. And the same method could be applied to May, August, November and the whole year, as shown in Fig. 3.4.

In February, the MLO for most sea areas in the South China Sea is over 40%, and the maximum occurrence center is located at the Luzon Strait and the southeastern sea waters of the Indo-China Peninsula, which is up to 70–80%. In the Bay of Bengal, the MLO for the middle and the northern parts is within 20% and the maximum zone is from Sri Lanka to Nicobar Islands (30–40%). In the Arabian Sea, the high occurrences are found in the southwest area (over 30%, up to 80% at the center).

In May, in the South China Sea, the MLO is from 20 to 40% in the maximum zone distributed in the northern coastal areas and the eastern coastal areas of the Indo-China Peninsula, which is in an arc shape surrounding the peninsula. In the Bay of Bengal, for most sea areas the MLO is over 30% with the maximum zone distributed in the northwestern sea areas and the southeastern sea areas around Sri Lanka. In the Arabian Sea, high occurrences are found in the northern and western sea areas, which are over 50%.

In August, the MLO are above 30% for most sea areas in the South China Sea with maximum zone located at the southeastern sea areas around the Indo-China Peninsula (over 70%). The Thailand Gulf has relatively higher occurrences which are from 60 to 70%. In the Bay of Bengal, the MLO is over 60% for most sea areas, with maximum zone located at the southeastern area of Sri Lanka (>90%). It is noteworthy that there is a minimum zone in the northeastern coastal areas around

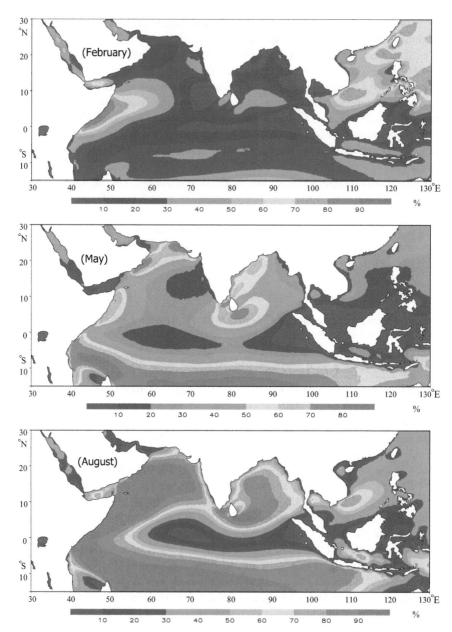

Fig. 3.4 Multi-year average occurrence of wind power density above 150 W/m² in February, May, August, November and annual mean value along the maritime silk road

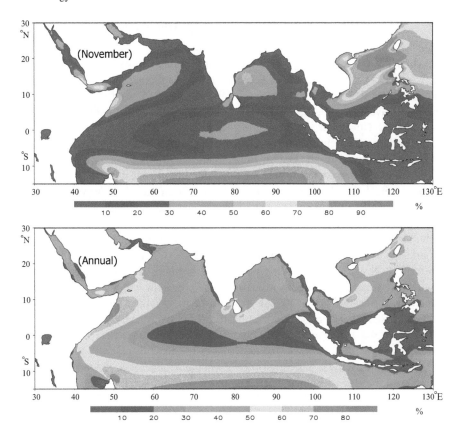

Fig. 3.4 (continued)

Sri Lanka as the southeastern monsoon is blocked by the island. In the Arabian Sea, occurrences in most sea areas are over 80% with a contour line of 90% covering most areas.

In November, the MLO is basically over 50% in the middle and the northern South China Sea, while the value of the Luzon Strait and its western parts could reach 90% and even above. In the Thailand Gulf and the sea waters of lower latitudes, it is generally less than 30%. In the Bay of Bengal, the maximum zone is located in the southwest with occurrences above 30% and the minimum zone located in the northeastern sea areas and the east of Sri Lanka.

Annual average, the MLO in the South China Sea and the North Indian Ocean is basically above 30% while the maximum zone is located from the Hainan Island to the Luzon Strait (50–60%), and the southeastern sea areas around the Indo-China Peninsula (50–70%). In the Bay of Bengal, the maximum zone is distributed at the southeastern part of Sri Lanka (50–60%) and the Manaar Gulf (50–70%). In the Arabian Sea, the maximum zone is located at the sea areas around Somali (50–80%).

3.4.3   Occurrences Wind Power Density Above 200 W/m²

Based on the 6-hourly wind power density in February in 1979, we counted the RLO on each 0.25° × 0.25°bin in February 1979. And under the same method, we calculated the RLO of each February from 1979 to 2015. And then obtained the multi-year average value of RLO in February. And the same could be applied to May, August, November and the whole year, as shown in Fig. 3.5.

In February, the South China Sea has relatively high RLO overall which is over 40% and in comparison the Bay of Bengal and the Arabian Sea have been clearly outshined in this perspective. There are three maximum zones: the southeastern sea waters of the Indo-China Peninsula, the Luzon Strait and the sea waters around Somali whose RLO is over 60%. The Manaar Sea and the southeastern sea areas have relatively high occurrences which are over 30%.

In May, as the cold air is drifting away, the RLO in the South China Sea is remarkably lower than that in February—less than 10% for the sea waters in the south of 10°N and 10–20% in the middle and northern parts of the South China Sea. During summer, the southwestern monsoon prevails in the North Indian Ocean. Generally, the value is over 20% in the west of the Arabian Sea and the most of the Bay of Bengal. The maximum zoon is located at the Manaar Sea (60–70%), the southeastern sea waters around Sri Lanka (60–80%) and sea waters around Somali (50–70%).

In August, the RLO is basically over 90% in the Arabian Sea, above 50% in most sea areas in the Bay of Bengal, over 20% in most of the South China Sea. And the percentages in the tropical South Indian Ocean are generally considerable.

In November, as the southwest monsoon fades away, the RLO in the North Indian Ocean drops to 40% and less rapidly. In the South China Sea, the RLO is obviously higher than that in the North Indian and the occurrences are over 50% in the middle and northern part, as the freezing air becomes dominant.

Annual mean value, for most areas along the Maritime Silk Road, the RLO is over 20% and there are three major maximum centers: the sea areas around Somali (over 50, 60% at the center), the southeastern sea areas around the Indo-China Peninsula (over 50%) and the Luzon Strait (over 50%). The Manaar Sea and the southeastern sea areas have relatively high occurrences which are over 40%.

3.4.4   Occurrences Wind Power Density Above 300 W/m²

Based on the 6-hourly wind power density in February in 1979, we found the ELO on each 0.25° × 0.25°grid point. And under the same method, we calculated the ELO of each February from 1979 to 2015 and its multi-year average value. And the same could be applied to May, August, November and the whole year, as shown in Fig. 3.6.

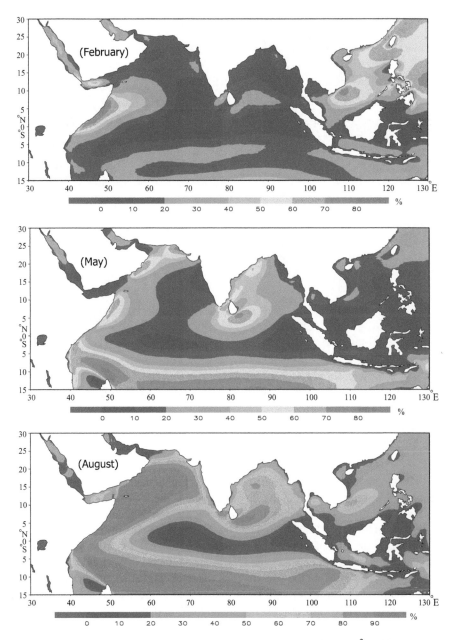

Fig. 3.5 Multi-year average occurrence of wind power density above 200 W/m² in February, May, August, November and annual mean value along the maritime silk road (after Zheng 2018b)

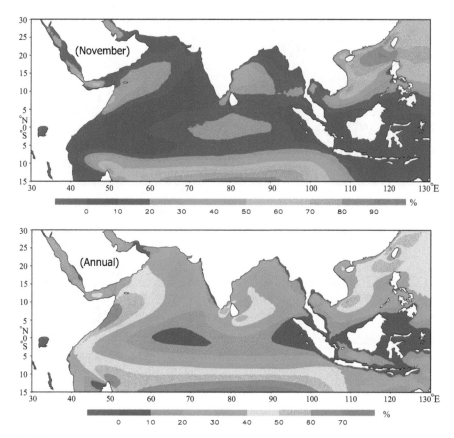

Fig. 3.5 (continued)

In February, the ELO of the South China Sea is higher than that of the Arabian Sea generally speaking, and that in the Bay of Bengal is the lowest among the three. In the South China Sea, occurrences for most areas are over 20% and could reach 55% in the Taiwan Strait, the Luzon Strait and the southeast sea waters around Indo-China Peninsula. In the Bay of Bengal, the number is within 5% for most sea areas, and only the Manaar Sea and a small sea area in the southeast of Sri Lanka are relative large areas (over 20%). In the Arabian Sea, relatively higher occurrences are found in the southwest (over 15%) with its center located at the coastal areas of Somali (around 55%) and those of the vast sea areas in the northeast of the Arabian Sea are within 10%.

In May, the Bay of Bengal has the highest ELO, followed by the Arabian Sea and the South China Sea comes at the end. In the South China Sea, the number is within 10% for most sea areas. In the Bay of Bengal, it is above 20 with 50% at the center. In the Arabian Sea, the north and the west has it above 20% while the rest has it below 20%.

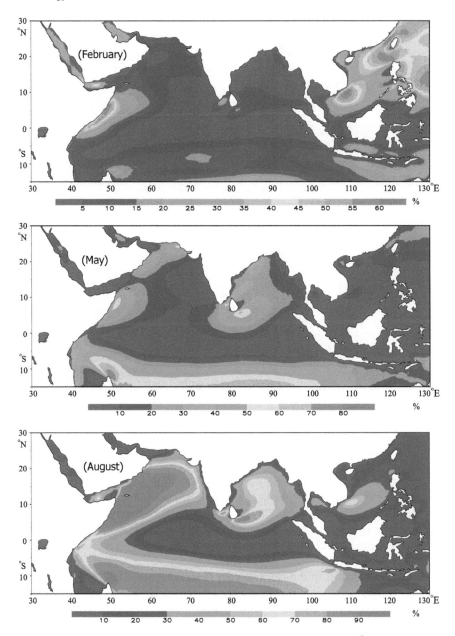

Fig. 3.6 Multi-year average occurrence of wind power density above 300 W/m^2 in February, May, August, November and annual mean value along the maritime silk road

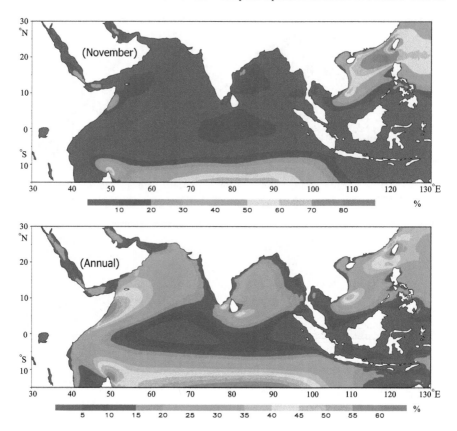

Fig. 3.6 (continued)

In August, the Arabian Sea has the highest ELO, followed by the Bay of Bengal and the South China Sea has the smallest. In the South China Sea, the ELO in the middle of sea areas is above 30% while the number is relatively lower for the rest. In the Bay of Bengal, most sea areas have it above 30 and 80% at the center. In the Arabian Sea, the ELO of most sea areas is above 60% which is relatively lower than that of the coastal areas of the Indian Peninsula.

In November, most of the North Indian Ocean is below 10%. The ELO in the South China Sea is much higher than that in the North Indian Ocean. The values in the middle and north of the South China Sea are greater than 20%, with large value center in the Luzon Strait and adjacent waters (above 70%).

On annual average, the Arabian Sea has the highest ELO, followed by the South China Sea and the Bay of Bengal comes at the end. The number in the Arabian Sea is above 20% and around 55% at the center. The South China Sea and the Bay of Bengal both have it above 15% while the former has larger sea areas with occurrences 40% and above compared with the latter.

3.4.5   *Occurrences Wind Power Density Above 400 W/m²*

Based on the 6-hourly wind power density in February 1979, we found the SLO on each $0.25° \times 0.25°$grid point. And under the same method, we calculated the SLO of each February from 1979 to 2015 and its multi-year average value. And the same could be applied to May, August, November and the whole year, as shown in Fig. 3.7. No matter in the South China Sea or the North Indian Ocean, the SLO in every representative month is quite low overall. In February and November, the South China Sea has the highest SLO, followed by the Arabian Sea and the Bay of Bengal comes at last, while in May, the ranking is in the opposite order with the Bay of Bengal coming at the top this time. In August, the Arabian Sea has the highest SLO, followed by the Bay of Bengal and the South China Sea comes at the end.

In February, most sea areas in the South China has a SLO over 20% with the contour line running in a northeast-southwest direction. The maximum center is located at the Luzon Strait and its western sea areas, the southeastern sea areas of the Indo-China Peninsula with occurrences up to 40%. The Beibu Gulf has relatively higher occurrences accounting for 15–25%. The only maximum center is located at the southeastern sea waters of Sri Lanka (5–10%). In the Arabian Sea, the SLO in most sea areas is less than 5%, while the maximum zone is located at the coastal areas around Somali (20–40%). It is noteworthy that the Manaar Gulf has relatively higher occurrences which are from 10 to 20%.

In May, the Bay of Bengal has the highest SLO and the sea areas with occurrences above 15% are obviously larger than those in the Arabian Sea and the South China Sea. And in the Bay of Bengal, the SLO of the middle and the western parts is higher than that of the east while that of the south is higher than that of the north. The maximum zone, which is in an oval shape, is distributed at the southeastern sea areas of Sri Lanka with occurrences up to 50% at the center. The relatively higher occurrences are found from the Manaar Gulf to the Palk Strait, which are around 25%. In the Arabian Sea, the maximum zone is distributed at the western coastal areas, especially those around Somali (20–45%). The number of the South China Sea is basically within 10%.

In August, in the Arabian Sea the SLO for most sea areas is above 60% and even 90% for the vast sea areas in the west. The contour line is running in a northeast-southwest direction. In the Bay of Bengal, the contour line forms a circle at the inlet of the bay with occurrences reaching 50% to 60% at the center. In the south, the contour line is running in a northeast-southwest direction with its center close to the northern coastal areas of Sri Lanka (over 60%). In the South China Sea, the SLO of the southeastern sea waters around the Indo-China Peninsula is from 30 to 50% while that of the rest within 20%.

In November, the SLO of the whole North Indian Ocean is within 10%. For vast sea areas in the middle and the north of the South China Sea, the number is over 20% with the maximum center located at the Luzon Strait and its western sea areas (over 60%). Relatively higher occurrences are found in the Beibu Gulf which are

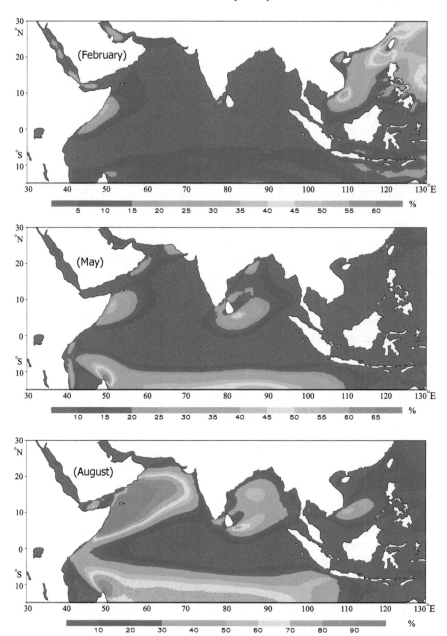

Fig. 3.7 Multi-year average occurrence of wind power density above 400 W/m^2 in February, May, August, November and annual mean value along the maritime silk road

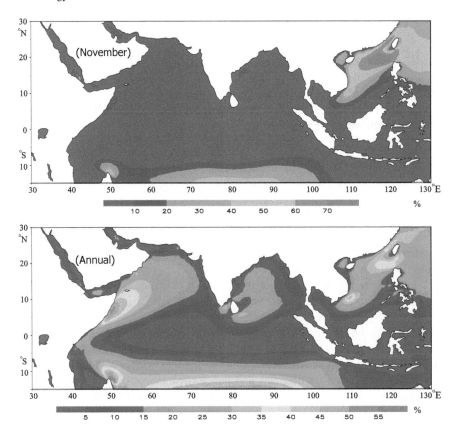

Fig. 3.7   (continued)

from 20 to 40% and those of the Thailand Gulf and sea areas of lower latitude are within 10%.

Annual occurrence: If we compare the spatial distribution of the annual value with that of each season, it is obvious that the annual average SLO of the Arabian Sea and the Bay of Bengal is propelled by the southwestern monsoon and that of the South China Sea is driven by the cold airs. In the Arabian Sea, the SLO is over 20% for most sea areas with the maximum center located at the coastal areas of Somali (over 50%). In the Bay of Bengal, the SLO of vast sea areas in the middle and the north is from 10 to 20% with a strip-shaped maximum zone, running in ENE-WSW direction, located at the southeastern part of Sri Lanka (20–30%). In the South China Sea, the SLO is over 20% for vast sea areas in the middle and northern sea areas with the strip-shaped contour line running in a northeast-southwest direction. There are two maximum centers distributed at the Luzon Strait and the southeastern sea areas of the Indo-China Peninsula (over 35%).

3.5 Energy Stability

Based on the 6-hourly wind power density from 1979 to 2015, we could calculate the coefficient of variation (Cv) of wind power density in February, May, August and November, which demonstrates the stability of the wind power density in each season. The smaller the Cv is, the more stable the density. The calculation method of Cv is as follows:

$$C_v = \frac{S}{\bar{x}} \tag{3.8}$$

$$S = \sqrt{\frac{\sum_{i=1}^{n} x_i^2 - (\sum_{i=1}^{n} x_i^2)/n}{n-1}} \tag{3.9}$$

where C_v is the coefficient of variation; $\bar{x}$ is the average value of wind power density; and S is the standard deviation.

Based on the 6-hourly wind power density in February in 1979, we calculated the C_v in this month. And under the same method, the C_v in each February from 1979 to 2015 was obtained. Then the multi-year average C_v for the past decades was obtained. And the same could be applied for May, August, November and the whole year, as shown in Fig. 3.8. Overall, the C_v of November is higher than those of the other representative months which means that the wind power is the least stable in November, as the South China Sea and the North Indian Ocean are in the period of monsoon transition season and the wind speed is less stable. The sea waters of middle latitude have more stable wind power than those around equator. And the same rings true for February. Overall, the wind power is more stable in May than it is in February, and wind power in the South China Sea is less stable than that in the Arabian Sea and the Bay of Bengal. In August, as the southwestern monsoon is strong and stable in sea areas around Somali, the Arabian Sea has the most stable wind power, followed by the Bay of Bengal and the South China Sea comes at the end.

In February, the C_v in the eastern sea areas of Somali is quite low, basically within 0.8, with the contour line running in an ENE-WSW direction; relatively higher value is found in the northwestern coastal areas of the Arabian Sea, which is generally above 1.2. In the South China Sea and the Bay of Bengal, sea areas in 5°–10°N have relatively smaller variation and the maximum center is found at the top of the Bay of Bengal which is in a roughly circular shape (basically over 1.0).

In May, the maximum zone changed from the northwestern coastal areas to the center of the Arabian Sea. In the Bay of Bengal, the C_v is higher in the east and smaller in the west. Minimum zone is distributed at the surrounding water areas of Sri Lanka and the northwestern coastal areas of the Bay of Bengal (less than 0.9). In the South China Sea, the minimum zone is in an arc shape surrounding the Indo-China Peninsula.

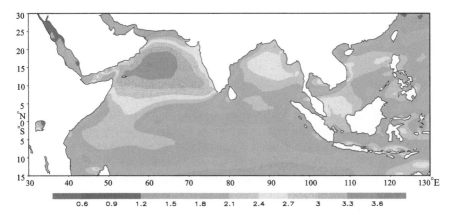

Fig. 3.9 Monthly variability index in February, May, August, November of the maritime silk road (after Zheng 2018b)

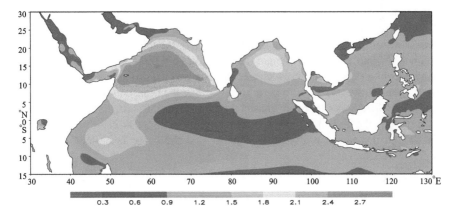

Fig. 3.10 Seasonal variability index in February, May, August, November of the maritime silk road

$$Mv = \frac{P_{M1} - P_{M12}}{P_{year}} \tag{3.10}$$

where P_{M1} is the wind power density in most abundant month, P_{M12} is the wind power density in the poorest month, P_{year} is annual mean wind power density.

$$Sv = \frac{P_{S1} - P_{S4}}{P_{year}} \tag{3.11}$$

where P_{S1} is the wind power density in most abundant season, P_{S4} is the wind power density in the poorest season, P_{year} is annual mean wind power density.

Averaging the wind power density from 00:00 on January 1st 1979 to 18:00 on January 31st 1979, we calculated the mean wind power density of January 1979.

Under the same method, we calculated the mean wind power density of each month from 1979 to 2014. Based on the monthly wind power density from January 1979 to December 1979 and combined with the equation, we calculated the Mv of the wind power density in 1979. And then adopting the same method, we calculated the Mv of each year from 1979 to 2014 and then obtained the multi-year average value of Mv, as shown in Fig. 3.9.

Averaging the wind power density from 00:00 on March 1st 1979 to 18:00 on May 31st 1979, we calculated the mean wind power density of Spring in 1979 (March–April-May). Under the same method, we calculated the wind power density in each season from 1979 to 2015. Based on the seasonal wind power density in 1979 and combined with the equation, we calculated the Sv of the wind power density in 1979. And then adopting the same method, we calculated the Sv in each year from 1979 to 2015 and the then obtained the multi-year average value of Sv, as shown in Fig. 3.10.

Mv: As shown in Fig. 3.9, the Arabian Sea has the largest Mv, followed by the Bay of Bengal and the South China Sea comes at the end. The Mv of the Arabian Sea is basically over 2.4 and 3.3 at the center. The Mv in the middle and the eastern part of the Bay of Bengal is from 2.1 to 2.7, which is obviously higher than that of the western coastal areas. In the South China Sea, the Mv of northern coastal areas is within 2.1 which is much lower than that of the middle and the south. In the Arabian Sea and the Bay of Bengal, the southwestern monsoon is clearly stronger than the cold airs, a situation which therefore caused the differences of Mv of these two regions. However, the cold air is stronger than the monsoon in the South China Sea, so the differences there are caused by the cold air instead.

Sv: From the Fig. 3.10, the spatial distribution of Sv is close to that of the Mv though the values are basically smaller and the cause is basically the same as mentioned above.

3.7 Energy Storage

The energy storage is closely linked to the volume of electricity generated. Previous researchers have tremendous work in this perspective but mostly on the storage in total. With reference to the method proposed by Zheng et al. (2012, 2013, 2018b), we calculated the wind power storage, including the total storage, exploitable storage and technological storage on each $0.25° \times 0.25°$ grid point. The equations are as follows:

$$E_{PT} = \bar{P} * H \tag{3.12}$$

$$E_{PE} = \bar{P} * H_E \tag{3.13}$$

$$E_{PD} = E_{PE} * C_e \tag{3.14}$$

where E_{PT} is the total storage of wind energy, $\bar{P}$ is the annual average wind power density, and H = 365d × 24 h = 8760 h, E_{PE} is the exploitable storage of wind energy and H_E is the annual total number of hours of effective wind speed, E_{PD} is the technological development volume of wind/wave energy resources. When about wind energy development, Ce = 0.785; i.e., the actual swept area of the wind turbine blades is 0.785, meaning that for a diameter of 1 m, the swept area of the wind turbine is calculated from 0.52 × p, which equals 0.785 m^2.

Based on the wind power density and the EWSO, we can calculate the wind power storage on each 0.25° × 0.25° grid point, as shown in Fig. 3.11. Generally, the spatial distribution of exploitable storage is in line with that of total storage as the EWSO is high overall along the Maritime Silk Road. We found out that the technological storage is only 78.5% of the exploitable storage based on the equation. In the Arabian Sea, the technological storage is relatively high which is over 0.8 × 10^3 kW·h·m^{-2} basically. The value in the sea waters around Somali is 2.4 × 10^3 kW·h·m^{-2} followed by the South China Sea with over 0.4 × 10^3 kW·h·m^{-2} and over 2.0 × 10^3 kW·h·m^{-2} in the middle part. In the sea areas around equator and in the Thailand Bay is within 0.4 × 10^3 kW·h·m^{-2} and the technological storage in the Bay of Bengal is obviously lower than that in the Arabian Sea and the South China Sea, which is basically above 0.4 × 10^3 kW·h·m^{-2} and the maximum zone is the Manaar Sea (above 1.2 × 10^3 kW·h·m^{-2}) and the southeastern sea areas of Sri Lanka (above 1.6 × 10^3 kW·h·m^{-2}).

3.8 Summary

In this chapter, we carried out the detailed research on the climate features of wind power along the Maritime Silk Road based on the ERA-Interim data, systematically covering the spatial and temporal distribution of a series of key indicators such as wind power density, EWSO, energy level occurrences, stability and storage of resources. The results are as follows:

(1) The wind power density along the Maritime Silk Road is promising overall and shows clear regional and seasonal differences. As for regional differences: the mean wind power density is basically over 200 W/m^2 in the Arabic Sea, around 150–450 W/m^2 in the South China Sea and the lowest in the Bay of Bengal but still above 150 W/m^2. There are three major maximum zones: the sea areas around Somali (over 400 W/m^2), the southeastern sea areas around Indo-China Peninsula (350 W/m^2) and the Luzon Strait (above 350 W/m^2). Relatively higher values are found in the Manaar Sea and the southeastern sea areas of Sri Lanka. As for seasonal differences: in the South China Sea, February and November have the highest wind power density, followed by August and May comes at the end, while in the Bay of Bengal, August has the highest, then May, followed by November and February. In the Arabian Sea, the wind power density is the largest in August, then in May and the smallest in February.

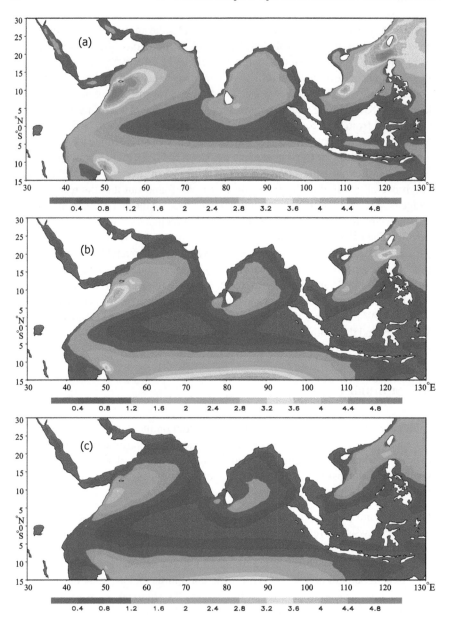

Fig. 3.11 Annual total storage **a** exploitable storage, **b** technological storage **c** of wind energy along the maritime silk road, unit: $\times 10^3$ kW·h·m^{-2}

(2) The regional differences in the wind power availability along the Maritime Silk Road: the EWSO is generally over 50% in the Arabian Sea, the Bay of Bengal and the South China Sea, which is good for wind exploitation. There are three major maximum centers: the sea areas around Somali (over 70%), the southeastern sea areas around the Indo-Peninsula (over 70%) and the Luzon Strait (over 70%). Relatively higher occurrences are found in the Manaar Sea which is between Sri Lanka and Indian Peninsula. As for seasonal differences: February and November have the highest availability in the South China Sea, followed by August and May comes at the end, while in the Bay of Bengal, August has the highest, then May and November, followed by February. In the Arabian Sea, the occurrences are the largest in August, then in May and the smallest in February.

(3) The ALO along the Maritime Silk Road is over 50% for most sea areas. In the Arabian Sea, the contour line goes in a northeast-southwest direction with the maximum center located at the sea areas of Sri Lanka (over 70%). In the Bay of Bengal, there are two relative maximum zones: the Manaar Gulf (over 60%) and the east of Sri Lanka which presents a strip-shaped area running in a southwest-northeast direction, where occurrences are over 60%. In the South China Sea, for most areas in the middle and northern parts, the occurrences are over 50% while the maximum zone is located at the southeastern sea areas of Indo-China Peninsula and the north of the South China Sea (over 60%). The RLO for most areas along the Road is over 20%/ and there are three major maximum centers: the sea areas around Somali (over 50%), the southeastern sea areas of the Indo-China Peninsula (over 50%) and the Luzon Strait (over 50%). Relatively high RLO values can be found in the Manaar Gulf.

(4) From the coefficient of variation (Cv), it is clear that the wind power is quite stable in the sea areas around Somali and Sri Lanka and the southeastern sea areas of Indo-China Peninsula in each season. From the Mv and the Sv, the wind power is less stable in the monthly and seasonal time scales in the Arabian Sea and the Bay of Bengal.

(5) The technological storage of wind energy of the Arabian Sea is basically over 0.8×10^3 kW·h·m^{-2} while that of the sea areas around Somali reaches 2.4×10^3 kW·h·m^{-2}. The amount of the South China Sea is smaller in comparison which exceeds 0.4×10^3 kW·h·m^{-2} for most sea areas. The amount of the Bay of Bengal is obviously lower than those of the Arabian Sea and the South China Sea.

In summary, the South China Sea and the North Indian Ocean are rich in wind power and suitable for wind power exploitation along the Maritime Silk Road. In the sea areas around Somali, the southeastern sea areas around the Indo-China Peninsula and the Luzon Strait, all the parameters such as the wind power density, the EWSO, the ELO, the energy storage and Cv indicate good quality of the wind power. And the same can be said of the parameters of the Manaar Sea and the southeastern sea areas of Sri Lanka though the Mv and the Sv of the wind power density are relatively high in these two areas.

References

Bao XH, Zhang FQ (2013) Evaluation of NCEP-CFSR, NCEP-NCAR, ERA-Interim, and ERA-40 reanalysis datasets against independent sounding observations over the Tibetan Plateau. J Clim 26:206–214

Capps SB, Zender CS (2010) Estimated global ocean wind power potential from QuikSCAT observations, accounting for turbine characteristics and siting. J Geophys Res 115:D09101. https://doi.org/10.1029/2009JD012679

Dee DP, Uppala SM, Simmons AJ et al (2011) The ERA-Interim reanalysis: configuration and performance of the data assimilation system. Q J R Meteorol Soc 137:553–597

Li Y, Wang Y, Chu HY, Tang JP (2008) The climate influence of anthropogenic land-use chanses on near-surface wind energy potential in China. Chin Sci Bull 53(18):2859–2866

Miao WW, Jia HJ, Wang D (2012) Active power regulation of wind power systems through demand response. Sci China Technol Sci 55:1667–1676

Zheng CW, Li CY (2015) Development of the islands and reefs in the South China Sea: wind climate and wave climate analysis. Periodical of Ocean University of China 45(9):1–6

Zheng CW, Li CY (2015) Development of the islands and reefs in the South China Sea: wind power and wave power generation. Periodical of Ocean University of China 45(9):7–14

Zheng CW, Li CY, Yang Y, Chen X (2016a) Analysis of wind energy resource in the Pakistan's Gwadar Port. Journal of Xiamen University (natural Science Edition) 55(2):210–215

Zheng CW, Li X, Chen X et al (2016b) Strategic of the 21st century Maritime Silk Road: Marine resources and development status. Ocean Develop Manage. 33(3):3–8

Zheng CW, Gao Y, Chen X (2017a) Climatic long term trend and prediction of the wind energy resource in the Gwadar Port. Acta Scientiarum Naturalium Universitatis Pekinensis 53(4):617–626

Zheng CW, Sun W, Chen X et al (2017b) Strategic of the 21st century Maritime Silk Road: Construction of Integrated Application Platform. Ocean Develop Manage. 34(2):52–57

Zheng CW, Li CY (2017) 21st century maritime silk road: big data construction of new marine resources: wave energy as a case study. Ocean Develop Manage 34(12):61–65

Zheng CW (2018) Energy predicament and countermeasure of key junctions of the 21st century maritime silk road. Pacific J 26(7):71–78

Zheng CW (2018) Wind energy evaluation of the 21st century maritime silk road. J Harbin Eng Univer 39(1):16–22

Zheng CW, Zhuang H, Li X, Li XQ (2012) Wind energy and wave energy resources assessment in the East China Sea and South China Sea. Sci China Technol Sci 55(1):163–173

Zheng CW, Pan J, Li JX (2013) Assessing the China Sea wind energy and wave energy resources from 1988 to 2009. Ocean Eng 65:39–48

Zheng CW, Li CY, Xu JJ (2019) Micro-scale classification of offshore wind energy resource—a case study of the New Zealand. J Clean Prod 226:133–141

Zheng CW, Zheng YY, Chen HC (2011) Research on wave energy resources in the northern South China Sea during recent 10 years using SWAN wave model. J Subtropical Res Environ 6(2):54–59. In Chinese with English abstract

Zheng CW, Li XY, Luo X, Chen X, Qian YH, Zhang ZH, Gao ZS, Du ZB, Gao YB, Chen YG (2019) Projection of future global offshore wind energy resources using CMIP data. Atmos Ocean 57(2):134–148

Zheng CW, Pan J (2014) Assessment of the global ocean wind energy resource. Renew Sustain Energy Rev 33:382–391

Zheng CW, Li XQ (2011) Wave energy resources assessment in the China Sea during the last 22 years by using WAVEWATCH-III wave model. Periodical of Ocean University of China 41(11):5–12. In Chinese with English abstract

Zheng CW (2020) Temporal-spatial characteristics dataset of offshore wind energy resource for the 21st Century Maritime Silk Road. China Scientific Data 5(4). (2020–12–22). https://doi.org/10.11922/csdata.2020.0097.zh

Chapter 4
Climatic Trend of Offshore Wind Energy in the Maritime Silk Road

Currently, the studies on the wind power along the Road area are still scarce, let alone those on its climatic trends. However, these studies are the theoretical basis which is crucial to improving the accuracy of the mid- and long-term forecast and could further impact the mid- and long-term planning of wind power exploitation (Zheng et al. 2016). Besides, nowadays there is abundant research on the evolution trends of the meteorology and ocean, but little on that of the wind power.

Zheng (2017a) groundbreakingly carried out studies on the trends of the offshore wind power and analyzed the overall evolution trend, its regional and seasonal differences and the relevance between the offshore wind power and some key factors in the North Atlantic Ocean. The results show that from 1988 to 2011, the wind power density of the North Atlantic Ocean was found to be significantly increasing at a speed of 4.45 (W/m^2)/yr from 1988 to 2011. And the increasing trends in the middle and high latitudes are stronger than that in the low latitude waters, while the increasing trend in the west waters is stronger than that in the east waters. The strongest increasing trend area is found to be located in the south waters of the Greenland of 35 (W/m^2)/yr. The relationship between the wind energy in the Atlantic Ocean and contemporaneous North Atlantic Oscillation (NAO) index is not significant, while there is no significant correlation between wind energy in the Atlantic Ocean and contemporaneous nino3 index. However, correlation between the nino3 index and wind power density (lagging 3 months) is obvious.

Current studies on the climatic trends of the wind power are extremely scarce and mainly focuses on the changes of the wave power density, and have not realized the climatic trend analysis of the wind power along the Maritime Silk Road, and hence are not able to provide theoretical support to the mid- and long-term projection of the wind power. Zheng (2017b) pioneered the research on the long-term climatic trends of wind power in the Gwadar Port and carried out relevant forecasts and analyses which could provide scientific guidance to the mid- and long-term planning of wind power exploitation for the port. Zheng (2018) furthered calculated the long-term evolution trends of a series of key parameters of wind power along the Maritime Silk Road. The results show that from 1979 to 2015, the wind power is abundant

C. Zheng et al., *21st Century Maritime Silk Road: Wind Energy Resource Evaluation*, Springer Oceanography, https://doi.org/10.1007/978-981-16-4111-4_4

in areas along the Road. The research also indicates that besides the wind power density, relevant studies on the evolution trends should focus more on parameters like EWSO, energy level occurrence and the stability of resources.

This chapter used the ERA-Interim data from ECMWF and analysed the historical climatic trends of wind power along the Maritime Silk Road, systematically covering the trends of a series of key parameters such as wind power density, EWSO, energy level occurrence, the stability. The trend analysis includes the regional and seasonal differences of the trends, to provide theoretical support for the mid- and long-term forecast of offshore wind power.

4.1 Data and Methods

This chapter mainly calculated and revealed the long-term climatic trends characteristics of wind power along the Maritime Silk Road, to provide theoretical support for the mid—and long—term forecast of offshore wind power. The data used in this part is the ERA-Interim data from ECMWF. The ERA-Interim data covers a spatial range from 90°S to 90°N, 180°W to 180°E and a time period from January 1st, 1979 to now. The spatial resolution used in this study is 0.25° × 0.25° and the temporal definition of 6 h. According to the 6-hourly ERA-Interim data from 1979 to 2015, we calculated the 6-hourly wind power density for 37 years along the Road.

In the process of wind power exploitation, the wind power density is the crucial parameter to evaluate the amount of wind energy. It is recognized that the wind speed between 5 and 25 m/s, known as the effective wind speed, is the optimal speed for collection and transformation of wind power, and the effective speed occurrence (EWSO) is closely linked to the availability of wind power. Generally, the wind power density over 200 W/m^2 (RLO) is considered to be high and RLO is a crucial indicator to evaluate richness of the wind power. The stability of wind power also has a direct impact on its collection, the efficiency of transformation and even the service life of facilities, and hence it is necessary to analyze the stability. Therefore, besides the wind power density, this chapter also has also taken parameters like EWSO, RLO and the stability, including Cv, Mv and Sv, into account. Based on the data on wind power density and ocean surface wind speed over a period of 37 years, we calculated the long-term climatic trends of wind power parameters along the Road, including the annual trends and the regional and seasonal differences of the trends. All the trends in this book are the trends after 3-point moving average.

4.2 Trend in Wind Power Density

Based on the 6-hourly wind power density from 1979 to 2015, we calculated the annual trends of wind power density on each 0.25° × 0.25° grid point in February, May, August and November from the South China Sea to the North Indian Ocean.

We also calculated the mean wind power density of February in 1979 by averaging the values from 00:00 February 1st, 1979 to 18:00 February 28th, 1979, and under the same method, we computed the mean wind power density in each February over a period of 37 years (from 1979 to 2015) and the climatic trends of wind power density in February for the past 37 years is calculated by using linear regression. Under the same method, we came up with the values of wind power density after 3-point moving average in February, May, August and November and their climatic trends, as shown in Fig. 4.1. And the trends after 3-point moving average are presented in Fig. 4.2.

In February, there are no significant changes shown in the trends of wind power density for most areas of the South China Sea and the North Indian Ocean while the areas that exhibit a significant increase are mainly distributed in the sea areas around Sri Lanka and the western sea areas around Sumatra. It is noteworthy that the Manaar Sea, which is between Sri Lanka and the Indian Peninsula, has a strong increasing trend, which is $2 \sim 4$ (W/m^2)/yr.

In May, there is no significant changes shown in the trends of wind power density for most areas of the South China Sea and the North Indian Ocean; the areas that exhibit a significant increase are mainly distributed in the coastal areas of Somalia (at a speed of 2–10 (W/m^2)/yr), the northern Arabian Sea ($0 \sim 4$ (W/m^2)/yr), most of the Bay of Bengal ($2 \sim 6$ (W/m^2)/yr), around the equator of the south Indian Ocean ($0 \sim 4$ (W/m^2)/yr) while areas with a significant decreasing trend are distributed in the east of Madagascar at a speed of $-8 \sim -2$ (W/m^2)/yr.

In August, the wind power density exhibits a strong declining trend in large sea areas of the middle and southern parts of the Arabian Sea and the middle and southern parts of the Bay of Bengal with the declining rate reaching -8 to -5 (W/m^2)/yr. This means that in the past 37 years, the southwest monsoon in the above regions has been weakening while the wind power density increases at a speed of 0–3 (W/m^2)/yr in the tropical South Indian Ocean. There are no obvious changes in other sea areas.

In November, the wind power density exhibits a significant increasing trend in the tropical South Indian Ocean, the northeastern part of Cape Hafun, the northeastern part of the Arabian Sea and the top of the Bay of Bengal while the areas that show a significant decreasing trend are relatively small, mainly distributed in the coastal waters of Fujian and Guangdong and the surrounding sea areas of Luzon Strait.

Annual trend: The areas showing a significant increase trend are mainly distributed in the tropical South Indian Ocean, the sea waters around Somalia and the northeastern Bay of Bengal. The increasing trend at the center could reach a speed from 2 to 3 (W/m^2)/yr while the areas with a significant decreasing trend are smaller which are mainly distributed in the eastern and western coastal areas of the Indian Peninsula and sea areas at low latitudes of the South China Sea.

Comparing the monthly and annual climatic trends, it is obvious that the trends of wind power density in different areas are dominated by different seasons. The increase trend of wind power density in the South Indian Ocean is mainly dominated by November, that in the sea areas around Somali is dominated by May and November while the decrease trend in the western coastal areas of Indian Peninsula chiefly dominated by February and August.

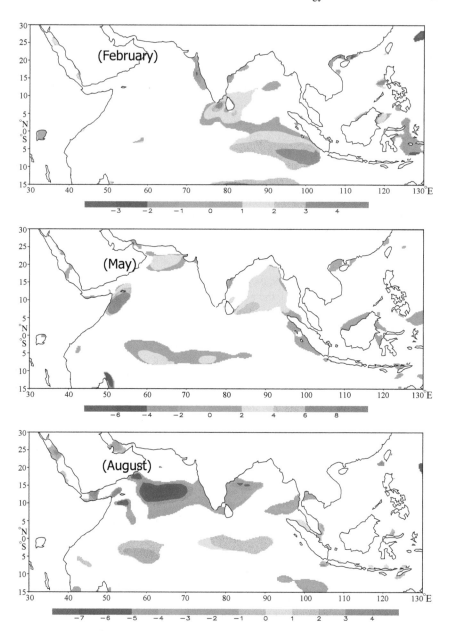

Fig. 4.1 Long-term trend of wind power density in February, May, August, November and annual average of the maritime silk road for the period 1979–2015 (after Zheng 2018), unit: $(W/m^2)/yr$. *Note* Only area significant at 95% level is coloured

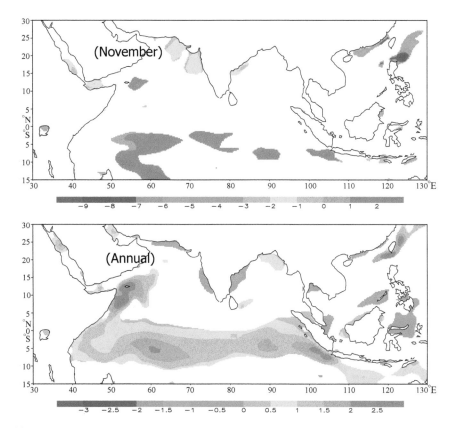

Fig. 4.1 (continued)

4.3 Trend in Effective Wind Speed Occurrence

Based on the 6-hourly wind field data from 1979 to 2015, we counted the EWSO of each month from January 1979 to December 2015, and then the climatic trends of values on each $0.25° × 0.25°$ grid point in the South China Sea and the North Indian Ocean in February, May, August, November and annual are calculated respectively during the past 37 years, as shown in Figs. 4.3 and 4.4. The areas with decreasing EWSO are not suitable for wind power exploitation and vice versa.

In February, there are no significant climatic variation in EWSO for most of the Arabian Sea, the top of the Bay of Bengal and most of the South China Sea. In the tropical sea areas of South Indian Ocean and the surrounding of Sri Lanka, the EWSO increases at a speed of 0.2–1.0%/yr (% hereby refers to EWSO instead of its variability, the same below), and the speed at the center could reach 1.0–1.2%/yr. The areas showing a significant decreasing trend are mainly distributed in the eastern and western coastal areas of the Indian Peninsula and the Beibu Gulf (at a speed of −1.2 to −0.4%/yr).

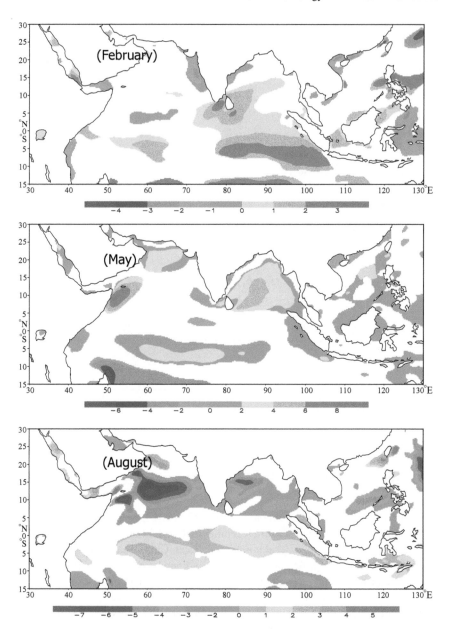

Fig. 4.2 Long-term trend of wind power density after 3-point moving average in February, May, August, November and annual average of the maritime silk road for the period 1979–2015, unit: $(W/m^2)/yr$. *Note* Only area significant at 95% level is coloured

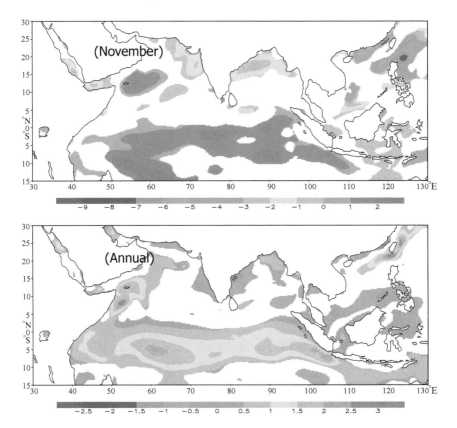

Fig. 4.2 (continued)

In May, the EWSO in most of the western Arabian Sea, most of the Bay of Bengal, and the traditional gale center of the South China Sea exhibits a significant increasing trend with a growth rate of 0.9%/yr in the maximum zone; only some scattered sea areas show a decreasing trend.

In August, there are no significant climatic variation in EWSO for most of the Arabian Sea, large sea areas around the Indo-China Peninsula and the Hainan Island; areas that exhibit a significant increasing trend are mainly distributed in a small region of the tropical areas of South India Ocean, which is in a striped shape running in a west–east direction. The areas with a significant declining trend are mainly distributed in most of the Bay of Bengal, the eastern and western coastal areas of the Indian Peninsula and the east of the South China Sea. It is noteworthy that when the wind power density in the Arabian Sea shows a strong downward trend, the EWSO exhibit no significant climatic trends.

In November, the areas that exhibit a significant increasing trend are mainly distributed in the tropical South Indian Ocean at a rate of 0.4–0.8%/yr and the north-eastern sea areas of Cape Hafun at a rate of 0.4–0.6%/yr while those that show a

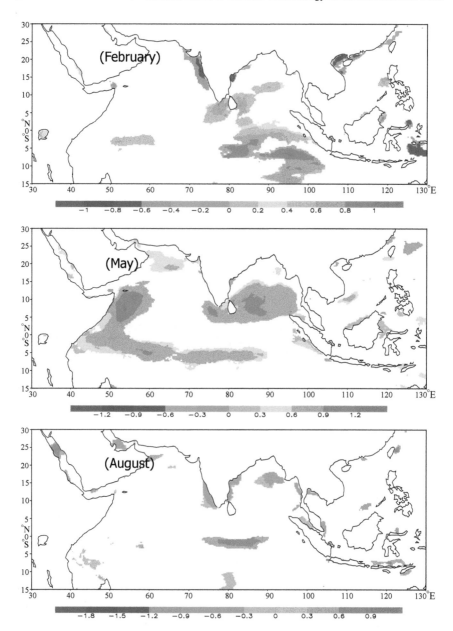

Fig. 4.3 Long-term trend of effective wind speed occurrence in February, May, August, November and annual average of the maritime silk road for the period 1979–2015 (after Zheng 2018), unit: %/yr. *Note* Only area significant at 95% level is coloured

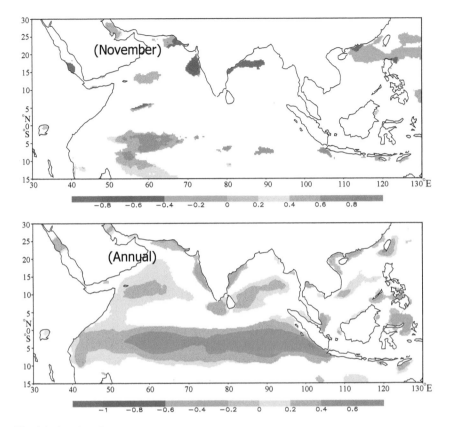

Fig. 4.3 (continued)

significant decreasing trend are mainly distributed in the northeastern part of the Arabian Sea (−1.0 to −0.2%/yr), the top of the Bay of Bengal (-1.0 to -0.4%/yr) and the northern part of the South China Sea (about −0.2%/yr).

Annual climatic trend: the EWSO in most sea areas of along the Maritime Silk Road has increased significantly year by year, especially in the tropical waters of the South Indian Ocean with a speed from 0.4 to 0.8%/yr. In comparison, areas with an annual declining trend are relatively small, mainly distributed at the eastern and western coastal areas of the Indian Peninsula, the northern coastal areas of the Arabian Sea and the surrounding waters of Hainan Island. Comparing the annual and monthly climatic trends, it is obvious that the EWSO in the South Indian Ocean is growing year by year in February, May and November, while the parameter of the eastern and western coastal areas of the Indian Peninsula is declining mainly in November and that of the surrounding sea areas of Hainan Island is mainly in February and November.

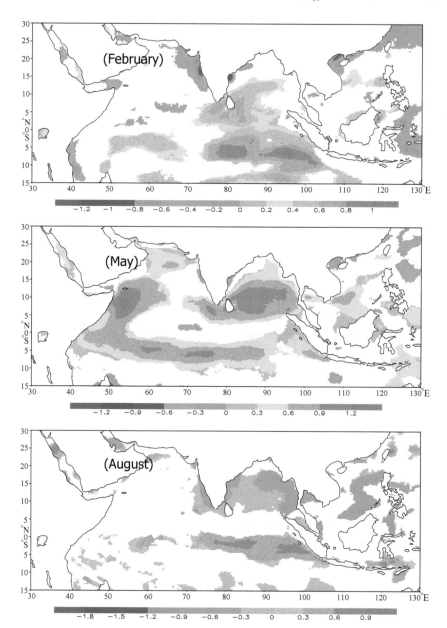

Fig. 4.4 Long-term trend of effective wind speed occurrence after 3-point moving average in February, May, August, November and annual average of the maritime silk road for the period 1979–2015, unit: %/yr. *Note* Only area significant at 95% level is coloured

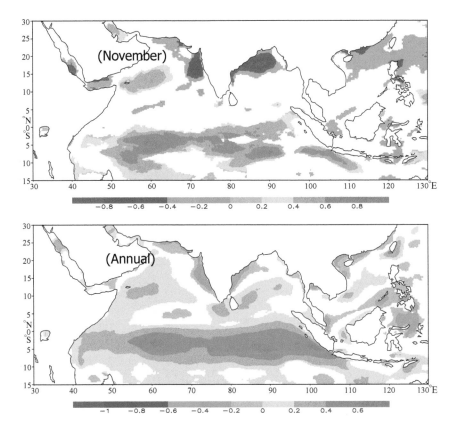

Fig. 4.4 (continued)

4.4 Trends in Energy Level Occurrences

Based on the 6-hourly wind field data from January, 1979 to December, 2015, we counted the rich level occurrences (RLO) of each month, and then the climatic trends of values on each $0.25° \times 0.25°$ grid point in the South China Sea and the North Indian Ocean in February, May, August and November respectively during the past 37 years, as shown in Figs. 4.5 and 4.6.

In February, in most sea areas of the Arabian Sea and the South China Sea and the top and east of the Bay of Bengal, there are no obvious variation in RLO. In the surrounding sea areas of Sri Lanka, the western sea areas of Sumatra, the values increase significantly at a speed of 0.2–0.8%/yr (% hereby refers to RLO instead of its variability, the same below) while only in scattered sea areas, the parameter exhibits a significant decreasing trend.

In May, in the middle and eastern Arabian Sea and the most of the South China Sea, there are no apparent changes in RLO. In the surrounding sea areas of Somali, the northern part of the Arabian Sea, the most of the Bay of Bengal and the middle

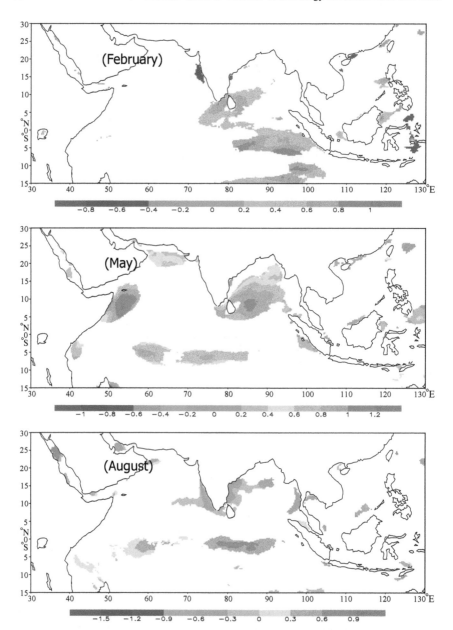

Fig. 4.5 Long-term trend of occurrence of wind power density greater than 200 W/m² in February, May, August, November and annual average of the Maritime Silk Road for the period 1979–2015 (after Zheng 2018), unit: %/yr. Note: Only area significant at 95% level is coloured

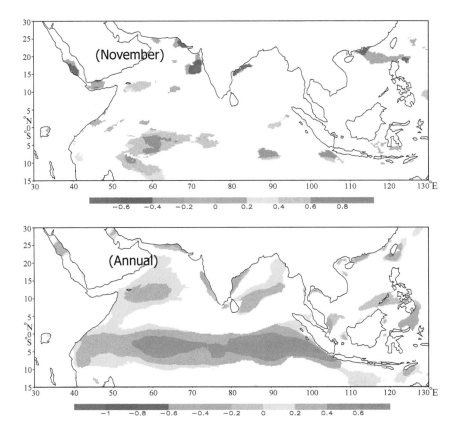

Fig. 4.5 (continued)

region of the equator south Indian Ocean, the value increases significantly at a speed of 0.4–1.4%/yr, 0.4–0.8%/yr, 0.4–1.2%/yr and 0.4–0.8%/yr respectively while only in some scatter sea areas, the parameter exhibits a significant decreasing trend.

In August, in the eastern and western sea areas of the Indian Peninsula which are in a vast stripped shape, the southeastern sea areas of the Bay of Bengal and the sea areas between the Indo-China Peninsula and Palawan, the RLO shows a marked decreasing trend while that in the tropical sea areas of the South Indian Ocean exhibits an opposite trend.

In November, in the top of the Bay of Bengal and the northern sea areas of the South China Sea, the RLO shows a significant decreasing trend; while in sea areas at low latitudes in the South Indian Ocean and the northeastern sea areas of the Ras Hafun, the parameter exhibits a strong growing trend.

Annual trend: in the tropical sea areas of South Indian Ocean, the western sea areas of the Arabian Sea and a narrow strip-shaped area running in a northeastern-southwestern direction in the Bay of Bengal, the RLO increases remarkably at a speed of 0.1–0.5, 0.1–0.3 and 0.1–0.2%/yr respectively. Those areas that show a

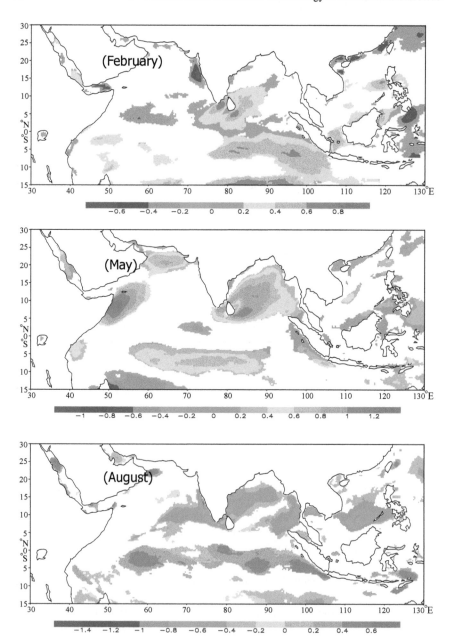

Fig. 4.6 Long-term trend of occurrence of wind power density greater than 200 W/m² after 3-point moving average in February, May, August, November and annual average of the Maritime Silk Road for the period 1979–2015, unit: %/yr. Note: Only area significant at 95% level is coloured

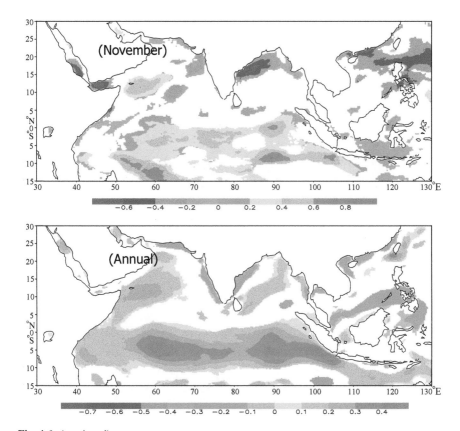

Fig. 4.6 (continued)

clear decreasing trend are relatively small, mainly distributed at the eastern and the western coastal areas of the Indian Peninsula, sea areas at the low latitudes of the South China Sea and the surrounding sea areas of Hainan Island. Comparing the monthly and annual climatic trends, it is clearly evident that the annual trend of RLO in the tropical South Indian Ocean can be found in each representative month while the declining trends mainly appear in August and February.

4.5   Trend in Energy Stability

Based on the 6-hourly wind power density from January 1979 to December 2015, we calculated coefficient of variation (Cv) in each month for the past decades, and then the climatic trends of Cv on each $0.25° \times 0.25°$ grid point in the South China Sea and the North Indian Ocean in February, May, August and November respectively during the past 37 years, as shown in Fig. 4.7. Besides, based on the Cornett's

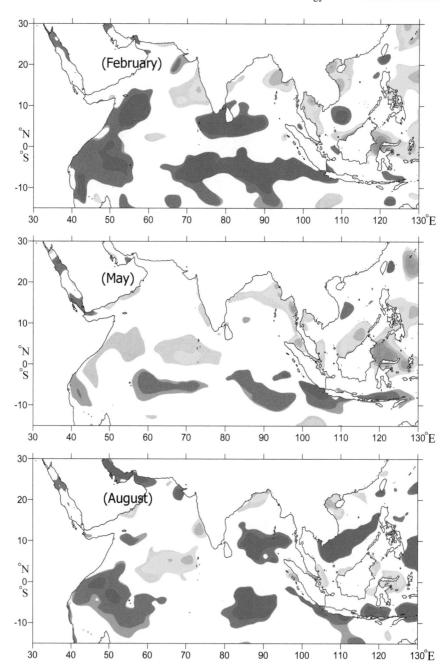

Fig. 4.7 Long-term trend of coefficient of variation after 3-point moving average in February, May, August and November of the maritime silk road for the period 1979–2015. *Note* Only area significant at 95% level is coloured

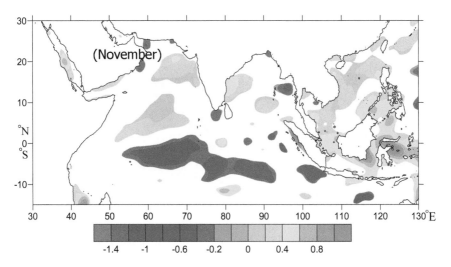

Fig. 4.7 (continued)

equation (Cornett 2008), we also calculated the Monthly Variability Index (Mv) and Seasonal Variability Index (Sv) of wind power in each year for the past decades and then their annual trends are calculated separately, as shown in Figs. 4.8 and 4.9. If the Cv indicates an increasing trend, it means that the stability of the resources is declining, and vice versa. The growing Mv and Sv values reflect that there are broader differences among months and seasons, a situation which is not suitable for wind power exploitation, and vice versa.

Climatic trend of Cv: in February, in the western and northern sea areas of the South China Sea, and eastern coastal areas of the Indian Peninsula, and northeastern

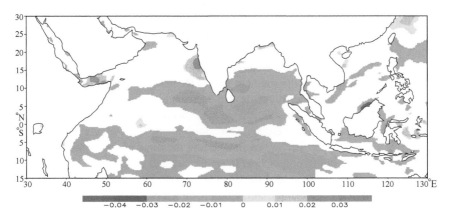

Fig. 4.8 Long-term annual trend of monthly variability index after 3-point moving average of the Maritime Silk Road for the period 1979–2015 (after Zheng 2018). *Note* Only area significant at 95% level is coloured

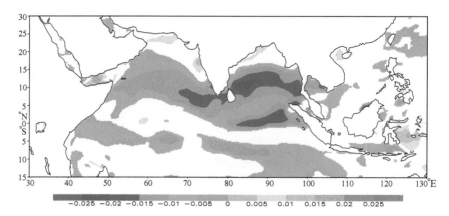

Fig. 4.9 Long-term annual trend of seasonal variability index after 3-point moving average of the Maritime Silk Road for the period 1979–2015 (after Zheng 2018). *Note* Only area significant at 95% level is coloured

and eastern part of the Arabian Sea, the growing values in Cv reflect a declining stability, while in the tropical sea waters of the South Indian Ocean and the Manaar Sea, the decreasing numbers in Cv show that the wind power is gaining more and more stability. In May, sea areas that have the marked declining Cv cover the vastest spaces among the four representative months, including large sea areas of eastern sea areas of Somali, the surrounding sea areas of Sri Lanka, the tropical sea areas of the South Indian Ocean and the southeastern sea areas of the Indo-China Peninsula. However, the values exhibit a significant increasing trend in the eastern and western coastal areas of Indian Peninsula, a small sea area in the northeastern Bay of Bengal and the surrounding sea areas of Indo-China Peninsula. There are no obvious climatic trends of Cv along the Maritime Silk Road in August and in November for most of the Arabian Sea, the middle and the north of the Bay of Bengal and the South China Sea. In the southeastern vast sea areas of the Indo-China Peninsula, the outlet of the Bay of Bengal and the southeastern sea areas of Somali, the parameter exhibits a strong decreasing trend.

Climatic trend of Mv: in the southeastern sea areas of the Arabian Sea, the vast sea areas of the middle and the southern Bay of Bengal and the tropical sea areas of the South Indian Ocean, the values show a decreasing trend, which means that the wind power density of each month tends to agree, a phenomenon which is beneficial to wind power exploitation. There is no obvious climatic trend of Mv in most of the Arabian Sea, the inlet of the Bay of Bengal and the most of the South China Sea while the declining values only appear in some scattered sea areas.

Climatic trend of Sv: generally, the Sv of wind power density along the Maritime Silk Road exhibit a decreasing trend which means the differences of wind power density in different seasons have been narrowing. This is good news for wind power exploitation. The declining values of Sv only appear in some scattered sea areas.

4.6 Summary

Based on the ERA-Interim data, we calculated the long-term climatic trends of wind energy resources along the Maritime Silk Road, including a series of key parameters of wind power density, EWSO, RLO and stability. And their annual trends, regional and seasonal differences of the trends are calculated. It is found that the offshore wind energy along the Maritime Silk Road has a positive climatic trend: the wind power density, EWSO and RLO in most areas exhibit an increasing or sable trends, meaning that the resource trends to be more abundant or stable. The Cv, Mv and Sv in most areas exhibit a decreasing or sable trends, meaning that the energy stability trends to be better or stable. And we came up with the following results:

(1) During the past 37 years from 1979 to 2015, the wind power density exhibits a year-by-year increasing trend in the tropical South Indian Ocean, sea areas around Somali, the northeastern sea areas of Bay of Bengal with the growth speed reaching 2–3 (W/m2)/yr at the center, while only in the eastern and western coastal areas of the Indian Peninsula and sea areas at low latitudes of the South China Sea the numbers show a decreasing trend. The wind power density of the tropical sea areas of the South Indian Ocean mainly grows in November while that of the waters of Somali mostly in May and November. The values in the eastern and western coastal areas of the Indian Peninsula that exhibit a decreasing trend mainly appear in February and August.

(2) For the past 37 years, the EWSO exhibit a significant year-by-year increasing trend for most sea areas along the Maritime Silk Road, especially in the tropical sea areas of the South Indian Ocean, with a growth rate of 0.4–0.8%/yr. Areas that exhibit a year-by-year decreasing trend are mainly distributed in the eastern and western coastal areas of the Indian Peninsula, the adjacent sea areas of the northern Arabian Sea and the small sea areas surrounding the Hainan Island with a decreasing rate of -0.6 to -0.2%/yr. The EWSO in the tropical waters of the South Indian Ocean mainly increases in February, May and November while that of the eastern and western coastal areas of Indian Peninsula decreases largely in November and that of the surrounding sea areas in Hainan Island generally decreases largely in February and November.

(3) During the past 37 years, the areas where the RLO has increased significantly are mainly distributed in the tropical waters of the South Indian Ocean (0.1 ~ 0.5%/yr), the western waters of the Arabian Sea (0.1 ~ 0.3%/yr), and a narrow strip-shaped sea area in the Bay of Bengal running in a northeast-southwest direction (0.1 ~ 0.2%/yr) while the declining values only appear in some scattered sea areas. The RLO in the tropical sea areas in the South Indian Ocean is growing throughout the year, while the declining trends of the eastern and western coastal areas of Indian Peninsula mainly appear in August and February.

(4) During the past 37 years, the Cv of wind energy density along the Maritime Silk Road has shown a significant decreasing trend or no obvious trends in each month while in a wide range of sea areas, the Mv and the Sv have also exhibit a

similar trend, and only in some scattered sea areas appears an increasing trend, a situation which is beneficial to wind energy exploitation.

References

Cornett AM (2008) A global wave energy resource assessment. In: Proceedings of the eighteenth international offshore and polar engineering conference. Vancouver, 2008: Paper no. ISOPE-2008-TPC-579

Zheng CW, Li CY, Li X (2017a) Recent decadal trend in the north atlantic wind energy resources. Adv Meteorol 7257492:8. https://doi.org/10.1155/2017/7257492

Zheng CW, Gao Y, Chen X (2017b) Climatic long term trend and prediction of the wind energy resource in the Gwadar Port. Acta Scientiarum Naturalium Universitatis Pekinensis 53(4):617–626

Zheng CW, Li CY, Pan J et al (2016) An overview of global ocean wind energy resources evaluation. Renew Sustain Energy Rev 53:1240–1251

Zheng CW (2018) Wind energy trend in the 21st century maritime silk road. J Harbin Eng Univer 39(3):399–405

Chapter 5
An All-Elements Short-Term Forecasting of Offshore Wind Energy Resource

The short-term wind power forecasting is closely linked to the collection and transformation of resources and their applications such as unit scheduling and power trading. In the early 1990s, some European countries have already been developing the wind power forecasting system and applying it into forecasting services (Lars and Watson 1994). In terms of forecasting technology, mid-term weather forecasts with the High-Resolution Limited Area Model (or High-Resolution Local Area Model) and nested inside to predict the amount of power generated on the fields. For example, the Denmark forecasting system, the Predictor, is currently used in Denmark, Spain, Ireland and Germany for short-term wind power forecasting services, while the Wind Power Prediction Tool (WPPT) is also applied in relevant services in Europe. Since the mid-1990s, the American company, True Wind Solutions, has also commercialized the wind power forecasting services. And the forecasting software that they have developed, the eWind, is a system comprised of mesoscale meteorological modeling and statistical modeling, used for forecasting the amount of power generated and other parameters on wind fields. Nowadays, the eWind and the Predictor are simultaneously working for two large wind power fields in California, the U.S. (Lars 1999; Bailey et al. 1999; Milligan et al. 2003; EPRI 2003). In October 2002, European Unions funded and initiated a plan called ANEMOS, aiming to optimize the forecasting model of the day which emphasized more on forecasting wind power over complex terrain and under extreme weather, as well as developing offshore wind power forecasting (Focken et al. 2001; Kariniotakis et al. 2003). The Canadian wind power numerical evaluation and forecasting software, the WEST, combined the mesoscale meteorological modeling (MC2) and WAsP, and it produced a wind atlas with a resolution of 100 to 200 m (Pinard et al. 2005; Nielsen et al. 2006). Other forecasting systems in use nowadays include the Previnento from German, the LocalPred and RegioPred from Spain and the HIRPOM from Ireland and Denmark etc. (Focken et al. 2001; Yu et al. 2006).

Previous researchers have done tremendous work on the short-term forecasting of offshore wind power across the globe. However, there are no such studies specifically targeting the countries and regions along the Maritime Silk Road. Zheng et al.

C. Zheng et al., *21st Century Maritime Silk Road: Wind Energy Resource Evaluation*, Springer Oceanography, https://doi.org/10.1007/978-981-16-4111-4_5

(2014) has proposed an intensive wind power forecasting method which interprets and applies these products into short-term forecasts and carried out case studies on forecasting the wind power during the interim between two cold air periods among China seas. Firstly, the precision of the forecasting wind field validated in Fig. 5.1 and Table 5.1. And then carried out the wind energy forecast, as shown in Figs. 5.2 and 5.3. Zheng et al. (2016) pointed out that there are potentials to transform the simple weather forecasting to the joint forecasting of weather and wind power, which is a new technical approach to the short-term forecasting of wind power. Besides, current forecasting products mainly focuses on parameters such as wind speed and wind power density while key indicators like availability, energy level occurrence, and storage amount have not been covered. In this chapter, we devised a set of wind power forecasting system for sea areas along the Road which comprised of more parameters and could better facilitate unit scheduling, power trading and management of wind field and power grid.

5.1   Methods

In this chapter, we devised an all-elements short-term wind power forecasting scheme for sea areas along the Road, which comprised of more parameters and could better facilitate unit scheduling, power trading, management of wind field and power grid. Zheng et al. (2019) pointed out that current forecasts are more often than not about wave power density while key parameters such as availability, energy level occurrences and storage amount should be given equal attention to, so as to exploit the resources to better serve their applications. To put it more specifically, we first calculate the hourly wind power density along the Road based on forecasts and equations of wind power density. Then we used hourly forecasts and the wind power density calculated to predict crucial parameters such as the availability (the occurrences of wind speed from 5 to 25 m/s), energy level occurrences (occurrences of winds with density over 100 W/m^2, 150 W/m^2, 200 W/m^2, 300 W/m^2, 400 W/m^2), energy storage (total storage, exploitable storage and technological storage), and carry out wind power forecasting which covers hourly wind power density, wind directions and wind energy roses.

In the early and mid-December 2016, a tropical cyclone appeared in the Bay of Bengal and at the same time, the cold air hit the South China Sea twice. At the end of December 2011, another strong tropical cyclone also has huge impact on the Bay. This chapter shall elaborate on the wind power forecasting of those two periods.

In this Chapter, a short-term forecasting of wind energy scheme for the next week is proposed. In the actual development of wind energy, the short-term forecasting of wind energy for the next days can be carried out by referring to this method.

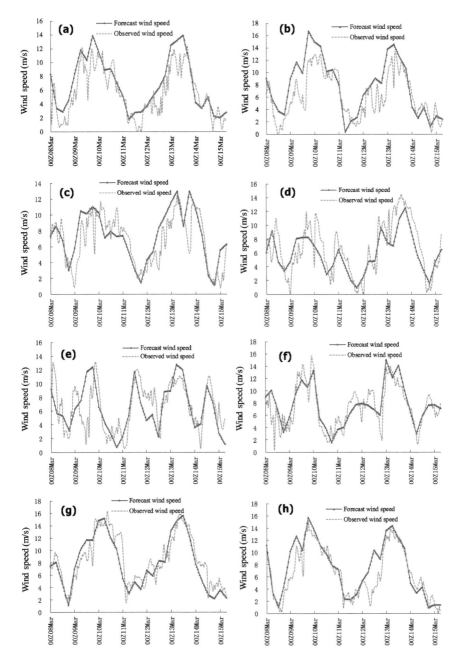

Fig. 5.1 Observed and simulated wind speed around the Korean Peninsula (after Zheng et al. 2014). **a–h** represent stations 22,101, 22,102, 22,103, 22,104, 22,105, 22,106, 22,107, 22,108 respectively

Table 5.1 Validation of the forecasting wind speed (after Zheng et al. 2014)

Station	Location		Correlation coefficient (CC)	Bias	Root mean square error (RMSE)	Mean absolute error (MAE)
	Longtitude (°E)	Latitude (°N)				
22,101	126.0	37.3	0.89	1.24	2.11	1.64
22,102	125.8	34.7	0.84	1.54	2.81	2.22
22,103	127.0	34.2	0.77	0.64	2.22	1.64
22,104	128.9	34.8	0.83	-0.70	2.23	1.84
22,105	130.0	37.5	0.66	-0.16	2.64	2.08
22,106	129.8	36.4	0.84	0.32	1.87	1.45
22,107	126.0	33.1	0.90	-0.45	1.84	1.49
22,108	125.8	36.3	0.90	0.89	2.11	1.46

5.2 Case Study One

5.2.1 Wind Field Forecast

In order to facilitate our observation, we mapped out the wind speed and direction along the Road based on wind field forecasts, using vector arrows to represent directions and different colors to distinguish speeds, as shown in Fig. 5.4.

At 0:00 on December 7th, the winds in the Bay of Bengal would create an almost closed circulation spinning anticlockwise, with the wind speed of the middle of the Bay exceeding 8 m/s. These are clear signs that a tropical cyclone is forming. At the same time, relatively strong cold air, blowing from the northeast, will be influencing the north of the South China Sea with the wind speed reaching 12 m/s and above. At 12:00 the tropical cyclone in the Bay of Bengal will be further forming with the wind swirling anticlockwise and here appears the eye of the tropical cyclone, a more apparent minimum center for wind speed. In the South China Sea, the freezing air slowly moves downwards the south. At 0:00 on December 8th, the trends would continue in the Bay of Bengal and at 12:00, in the Bay of Bengal the tropical cyclone intensifies, reaching a speed of 14 m/s at the center. The cold air in the South China Sea slowly goes south and the areas with wind speed up to 12 m/s and above are obviously shrinking, a phenomenon which means that the cold air begins weakening. At 0:00 on December 9th, the tropical cyclone would slowly move towards the northwest and weaken in the process while at 12:00 the cold air is about to exert its impact on the north of the South China Sea. At 0:00 on December 10th, the cyclone center would arrive at the middle of the Bay of Bengal and its momentum continue to decrease. Strong winds are found in the first, second and fourth quadrant. In the South China Sea, the second wave of cold air would bring high winds blowing at over 12 m/s to the sea areas from the Luzon Strait to the Hainan Island. At 12:00, in the Bay of Bengal the tropical cyclone would gain momentum again. However, during this time, the strong winds would be swirling at the front of the cyclone (in the first

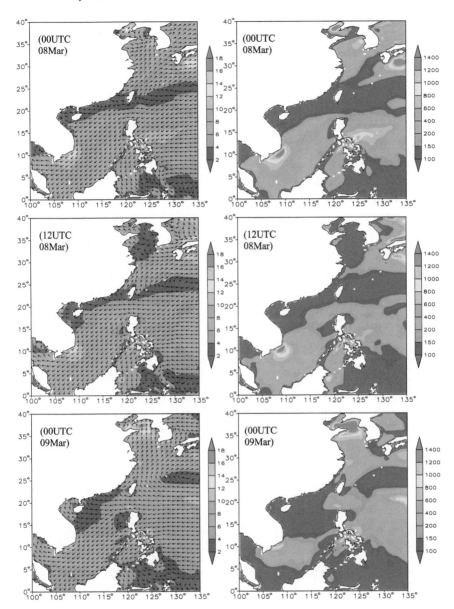

Fig. 5.2 Forecasting of wind field and wind power density for two cold air process in March 2013 in the China seas (after Zheng et al. 2014)

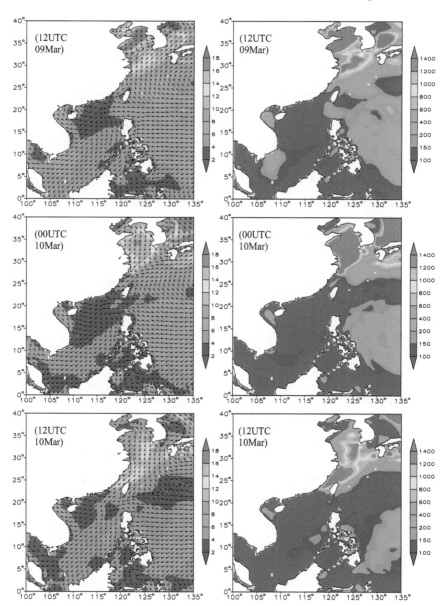

Fig. 5.2 (continued)

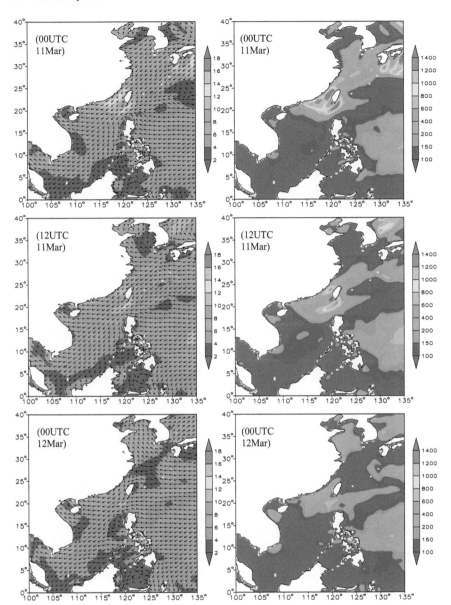

Fig. 5.2 (continued)

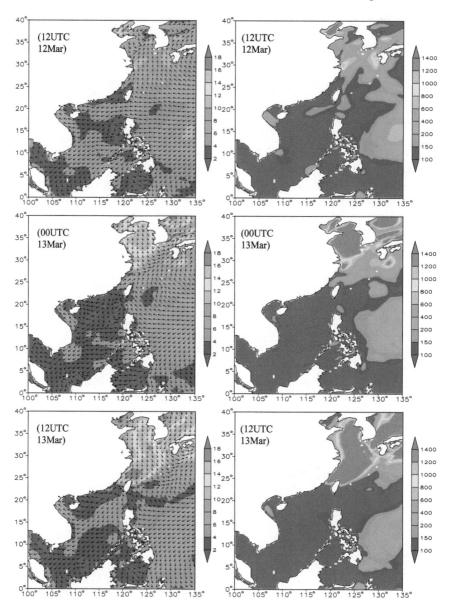

Fig. 5.2 (continued)

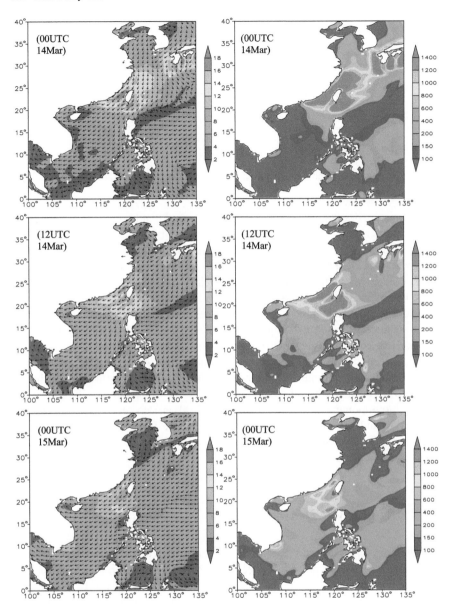

Fig. 5.2 (continued)

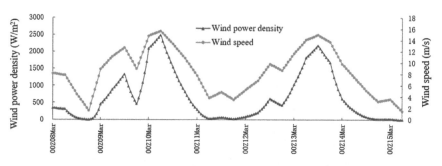

Fig. 5.3 Forecasting of wind speed and wind power density at one station (after Zheng et al. 2014)

and second quadrants), while notably it is recognized that high winds are generally found on the right side of the movement direction of the cyclone (in the first and fourth quadrants) in the traditional opinion. While the cyclone is moving towards the northwest, it is blocked by the landscape causing the energy to pile up. This means that key indicators such as terrain also have their impact on where the high winds are. Thus, hereby we further complete the distribution characteristics of the high winds: based on the movement direction of the cyclone, when it is at an open sea area, strong winds are mainly located at the right side (in the first and fourth quadrant); otherwise, the winds are generally found at the front (in the first and second quadrants) when the moving direction of cyclone was blocked by the terrain. At 0:00 on December 11th, the cyclone would continue to advance towards the northwest in the Bay of Bengal and keep weakening while the cold air in the South China Sea slowly goes south gaining more momentum in the process. At 12:00 on December 11th, the cyclone is about to land with the fastest winds blowing at the front. On December 12th, it lands on the Indian Peninsula, and the second wave of cold air influencing the South China Sea would phase away.

5.2.2 Wind Power Density Forecast

Based on the forecasts and the equations of the wind power density, we calculated the hourly wind power density in December 2016 (as shown in Fig. 5.5). At 0:00 on December 7th, the cold air has brought tremendous wind power to the north of the South China Sea with the wind power density basically over 1000 W/m^2 and 1600 W/m^2 at the center. The tropical cyclone is forming in the Bay of Bengal where the wind power density of most sea areas exceeds 200 W/m^2 and even 1200 W/m^2 at the center which is rather small in terms of space. At 12:00 on December 7th, the tropical cyclone becomes slightly stronger in the Bay of Bengal and the areas with wind power density over 1400 W/m^2 disappear as the cold air goes south in the South China Sea. At 0:00 on December 8th, the tropical cyclone would strengthen tremendously in the Bay and the maximum values of wind power density are found at

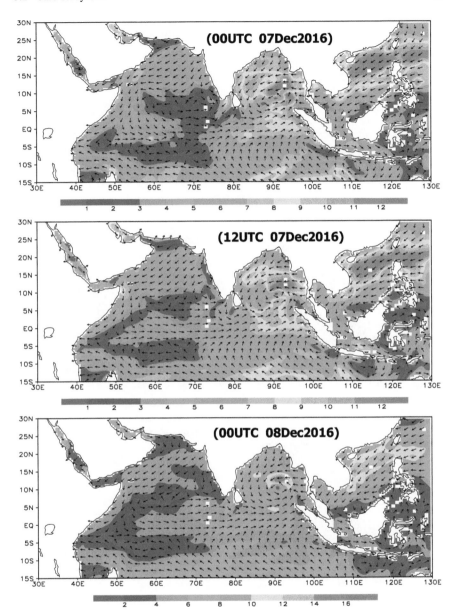

Fig. 5.4 Forecasting of wind field for the next week (start from 00:00 7th, December 2016) of the maritime silk road, unit of wind speed: m/s

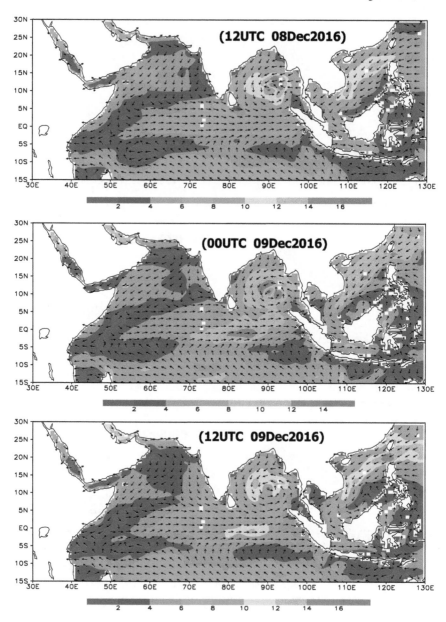

Fig. 5.4 (continued)

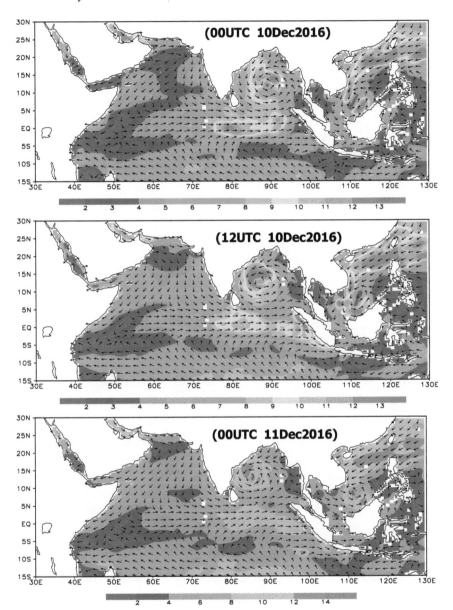

Fig. 5.4 (continued)

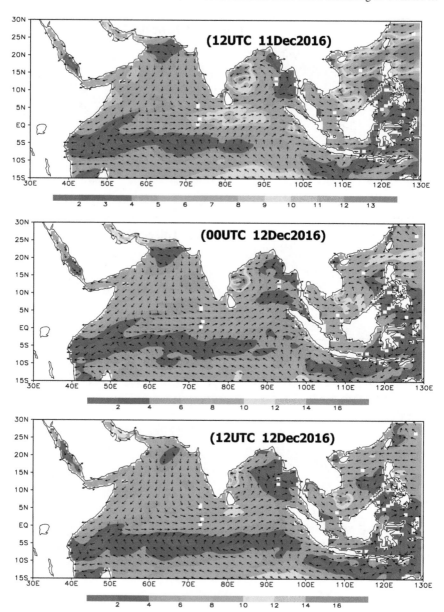

Fig. 5.4 (continued)

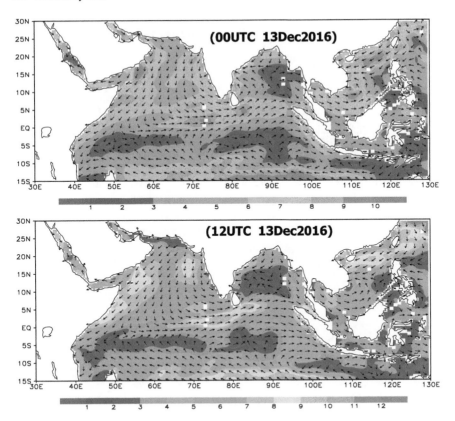

Fig. 5.4 (continued)

the right side of the cyclone's movement direction with values reaching 2700 W/m^2 at the center. In the South China Sea, sea areas which are significantly affected by the cold air have the wind power density ranging from 600 W/m^2 to 1800 W/m^2. At 12:00 on December 8th, in the Bay of Bengal, sea areas with wind power density over 900 W/m^2 expand drastically and in the South China Sea, the wind power density significantly drops as the cold wind going south gradually loses its momentum. From 0:00 on December 9th to 0:00 on 10th, the cyclone would form a more obvious circular spinning pattern and in the South China Sea, the first wave of cold air would continuously decrease while the second wave gradually gains more power. At 12:00 on December 10th, as the tropical cyclone strengthens again in the Bay of Bengal, the wind power density would also increase dramatically with the maximum density distributed at the front of the cyclone's movement direction (reaching 1600 W/m^2). Later, as the tropical cyclone slowly moves towards the northwest and lands on the Indian Peninsula, the wind power density gradually decrease. It is obvious that the constant and strong cold air would bring rich wind power to the South China Sea.

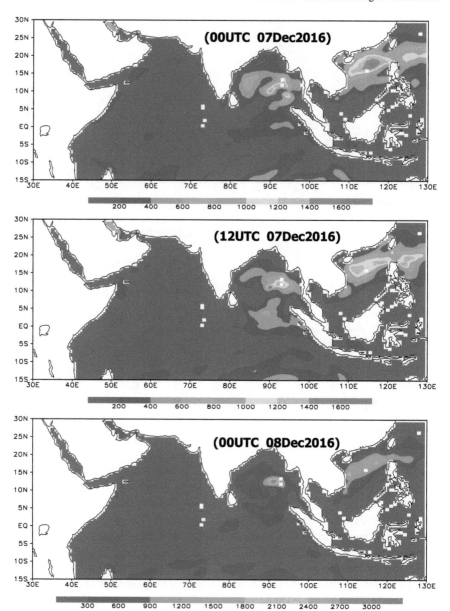

Fig. 5.5 Forecasting of wind power density for the next week (start from 00:00 7th, December 2016) of the maritime silk road, unit: W/m^2

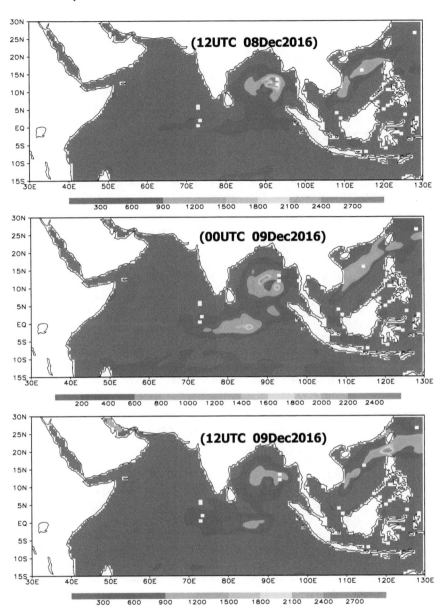

Fig. 5.5 (continued)

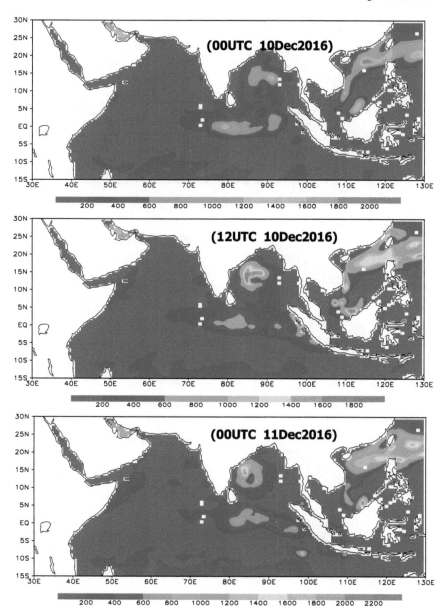

Fig. 5.5 (continued)

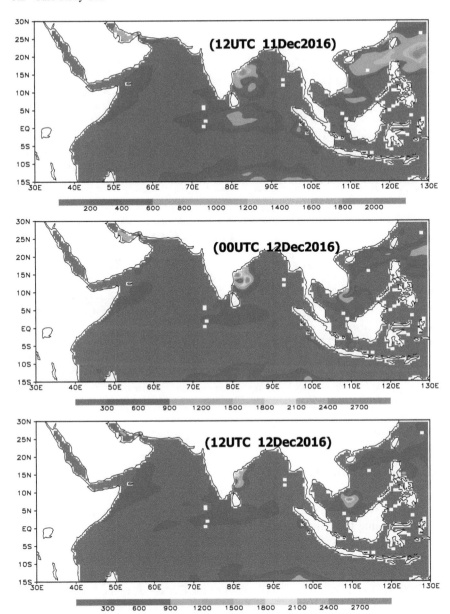

Fig. 5.5 (continued)

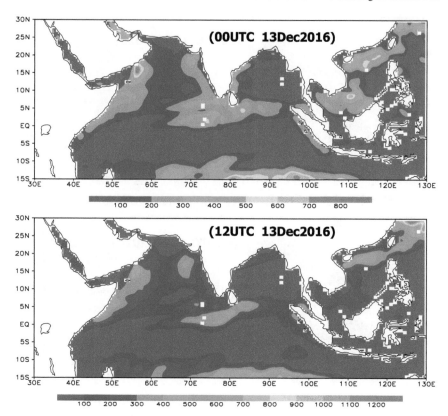

Fig. 5.5 (continued)

5.2.3 Availability Forecast

In Chap. 3, we pointed out that the availability of resources should be considered when analyzing the climate features; and similarly, the short-term availability is also crucial to the short-term wind power forecasts. Therefore, we will forecast the availability for the upcoming week. It can also be applied for ten to fifteen days based on specific needs. Based on the hourly wind field data from 0:00 on December 7th to 12:00 on December 13th, we will calculate the EWSO along the Road during this time period, so as to predict the availability of wind power in the upcoming week (as shown in Fig. 5.6).

Generally speaking, the wind power availability of the South China Sea and the North Indian Ocean looks promising in the following week. In the South China Sea, due to the strong and constant cold winds, the EWSO in most areas is over 90% while the values in the Beibu Gulf and the Bay of Thailand are relatively small as the landscape is blocking the cold air. Besides, the southeastern coastal areas of the South China Sea have relatively small EWSO as well. In the Bay of Bengal, the EWSO is

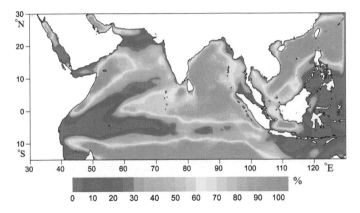

Fig. 5.6 Forecasting of wind energy availability (effective wind speed occurrence) for the next week (start from 00:00 7th, December 2016) of the Maritime Silk Road, unit: %

above 80% for most sea areas with the minimum zone located at the eastern coastal areas. In the Arabian Sea, the EWSO is obviously lower than that in the South China Sea and that in the Bay of Bengal. However, for most sea areas the value exceeds 40% with the maximum zone distributed at the coastal areas of Somali (over 90%) and the minimum zone at the northern tip of the Arabian Sea (basically less than 20%).

5.2.4 Energy Level Occurrences Forecast

In Chap. 3, we elaborated on the importance of the energy level occurrences and hereby we will propose the short-term forecasts on the energy level occurrences. Based on the hourly wind power density from 0:00 on December 7th to 12:00 on December 13th, we will come up with the energy level occurrences in the upcoming week, including occurrences of wind power density of 100 W/m^2, 150 W/m^2, 200 W/m^2, 300 W/m^2, and 400 W/m^2 (as shown in Fig. 5.7).

ALO: the values exceed 90% in most sea areas of the South China Sea, 60% in the Bay of Bengal, 30% in the Arabian Sea. In the sea areas above, the spatial distribution of ALO is basically in line with the that of EWSO predicted.

MLO: the values surpass 90% in most sea areas around the South China Sea, 50% in the Bay of Bengal and 30% only in sea areas surrounding Somali in the Arabian Sea.

RLO: the values are over 90% in the north of the South China Sea, 80% in the coastal areas southeast to the Indo-China Peninsula, 40% in most sea areas of the Bay of Bengal. In the surrounding sea areas of Somali, areas with occurrences over 30% are smaller than those of MLO.

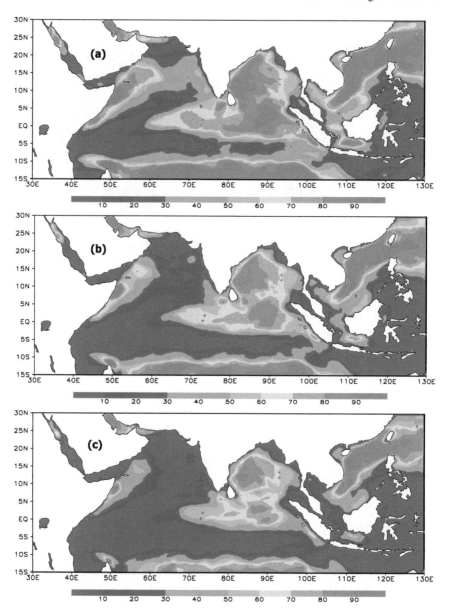

Fig. 5.7 Forecasting of energy level occurrences with wind power density above 100 W/m² **a**, 150 W/m² **b**, 200 W/m² **c**, 300 W/m² **d**, 400 W/m² **e** for the next week (start from 00:00 7th, December 2016) of the Maritime Silk Road, unit: %

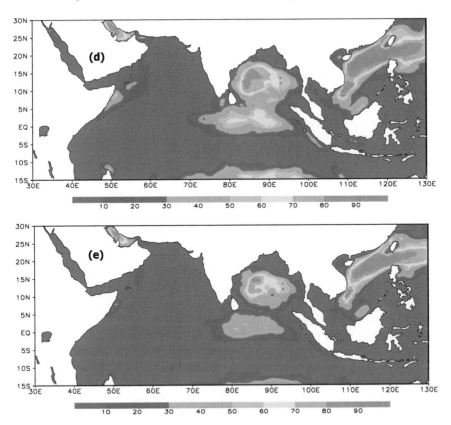

Fig. 5.7 (continued)

ELO: in the South China Sea, the relatively higher occurrences are found in sea areas along the Luzon Strait, the Hainan Island and the southeastern sea areas to the Indo-China Peninsula with ELO basically over 70% (and over 90% at the center). In the Bay of Bengal, the occurrences are above 40% in most sea areas, but less than 10% in the inlet of the Bay. In the Arabian Sea, the ELO is basically less than 10% for most sea areas and only in small sea areas around Somali, the values are up to 30%.

SLO: the relatively higher occurrences are found in sea areas along the Luzon Strait, the Hainan Island and the southeastern sea areas to the Indo-China Peninsula with SLO above 60%. The maximum zone is distributed at the Luzon Strait and its western sea areas with occurrences exceeding 90%. The SLO is basically within 10% and over 30% in the northern part and the middle and southern part of the Bay of Bengal respectively. The SLO in the Arabian Sea is generally with 10%.

5.2.5   Energy Storage Forecast

In Chap. 3, we pointed out why it's necessary to take the wind power storage of a climate into account and hereby we would analyze the short-term forecasts on this parameter in this chapter. Based on the wind power density and EWSO data from December 7th to 13th, we calculate the total storage, the exploitable storage and the technological storage. The equations are as follows:

$$E_{PT} = \overline{P} * H \tag{5.1}$$

$$E_{PE} = \overline{P} * H_E \tag{5.2}$$

$$E_{PD} = E_{PE} * C_e \tag{5.3}$$

where E_{PD} is the total storage of wind energy for the next week, $\overline{P}$ is the average wind power density for the next week, and $H = 7d \times 24\,h = 168\,h$, E_{PE} is the exploitable storage of wind energy for the next week and H_E is the total number of hours of effective wind speed for the next week, E_{PD} is the technological development volume of wind/wave energy resources for the next week. When about wind energy development, Ce $= 0.785$; i.e., the actual swept area of the wind turbine blades is 0.785, meaning that for a diameter of 1 m, the swept area of the wind turbine is calculated from $0.52 \times p$, which equals $0.785\,m^2$. In this Chapter, the wind energy scheme for the next week is proposed. In the actual development of wind energy, the wind energy storage for the next days can be carried out by referring to this method.

Total storage (as shown in Fig. 5.8a): in the upcoming week, the South China Sea has the largest wind power storage, followed by the Bay of Bengal, and the Arabian Sea has the lowest. In the South China Sea, the maximum zone is located along the Luzon Strait and its southwestern sea areas with the total amount exceeding $200\,kW \cdot h/m^2$. In the Bay of Bengal, the inlet has the lowest occurrences which are within $40\,kW \cdot h/m^2$ while the middle and the southern parts have relatively higher occurrences from 80 to $120\,kW \cdot h/m^2$. In the Arabian Sea, the total storage is less than $40\,kW \cdot h/m^2$ for most sea areas in the following week.

The exploitable storage (as shown in Fig. 5.8b) and the technological storage (as shown in Fig. 5.8c): the spatial distribution of these two parameters follows the same trend with that of the total storage amount only with slightly lower values, a phenomenon which is likely due to the relatively high EWSO along the Road during this period.

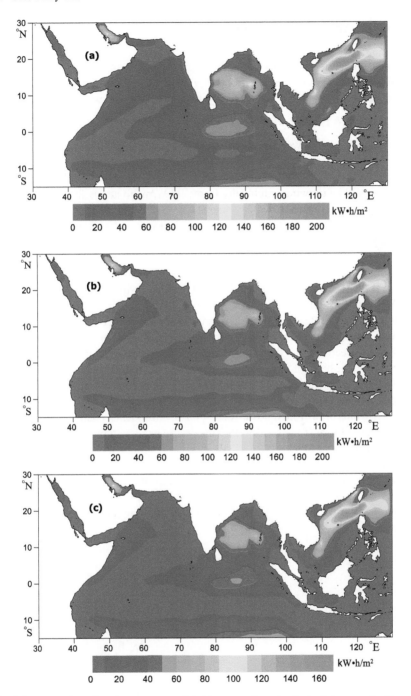

Fig. 5.8 Forecasting of total storage **a**, exploitable storage **b**, technological storage **c** for the next week (start from 00:00 7th, December 2016) of the maritime silk road

5.2.6 *Forecast of Key Point*

We randomly chose two sites out of three: the Arabian Sea, the Bay of Bengal and
the South China Sea (as shown in Fig. 5.9), and carried out wind power forecasts
including the hourly wind power density, wind direction and wind energy rose.

(1) **Hourly WPD and wind direction**

The hourly wind power density and the wind direction are shown in Fig. 5.10 and
Table 5.2

Site A (as shown in Fig. 5.10a): from 0:00 on December 7th, the wind power
density would be growing continually from 100 W/m² to nearly 300 W/m². During
the night of 7th, it begins to decrease and reaches its bottom at 0:00 on 9th. And then
the value varies slightly between 100 W/m² to 200 W/m² till 0:00 on 11th. However,
it rapidly increases to 300 W/m² to 400 W/m² which might be due to the influence
of a weak cold air mass. During the upcoming week, the wind direction at the Site
A maintains at an angle around 40 degrees.

Site B (as shown in Fig. 5.10b): the wind power density would be dropping from
170 W/m² to less than 30 W/m² since 0:00 on 7th. From the midnight of 7th to that
of the 10th, the wind comes from 30 degrees and then changed to 330 degrees.

Site C (as shown in Fig. 5.10c): from 0:00 on 7th to 0:00 on10th, the wind power
continues to decrease and then slowly varies. However, a rapid increase is observed
starting from the night of 11th. The wind at this site mainly comes from the southwest.

Site D (as shown in Fig. 5.10d): significantly influenced by the tropical cyclone,
the site has the wind power density above 400 W/m² in the following week. The
northerly winds blow from 0:00 on 7th to 0:00 on 9th and then change their directions
from the west towards the southwest. This can be explained by the fact that the site
D is under the front of the cyclone in the beginning and then covered by the back as
the cyclone moves towards the northwest.

Site E (as shown in Fig. 5.10e): under the control of a cold air mass, the wind
power density is basically within 100 W/m² and value would go a bit lower during
the interim of two masses. When the cold air mass is going south, the surface below it

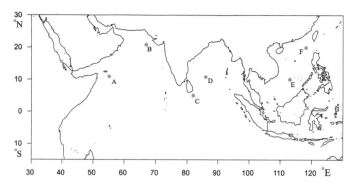

Fig. 5.9 Distribution of sites

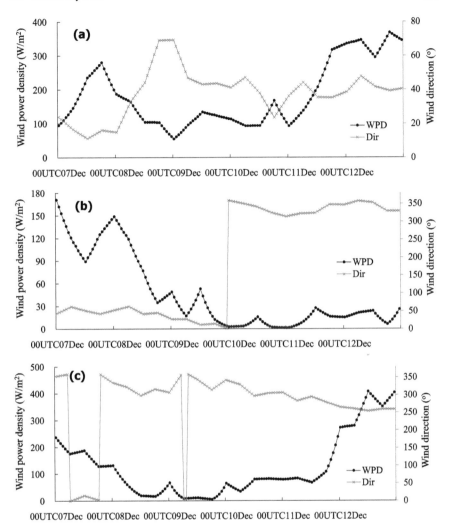

Fig. 5.10 Forecasting of hourly wind power density and wind direction at sites A (**a**), B (**b**), C (**c**), D (**d**), E (**e**), F (**f**) start from 00:00 7th, December 2016

becomes warmer and thus weakens the mass as it is at relatively low latitudes. When arriving at Site E, the mass would have lost most of its momentum compared with it at Site F and winds would be blowing mainly from the northeast in the upcoming week.

Site F (as shown in Fig. 5.10f): the site, located at the north of the South China Sea, is drastically influenced by the cold air and thus wind power density is basically over 800 W/m², even reaching 1600 W/m² at the peak in the following week. The wind steadily blows from the northeast at the site F during this time.

(2) **Wind energy rose**

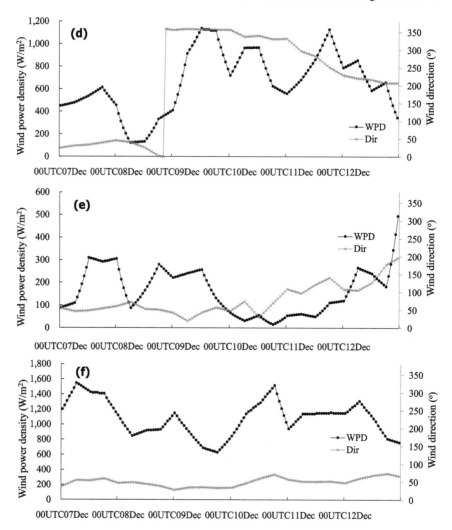

Fig. 5.10 (continued)

According to the data on the hourly wind power density and wind directions, we mapped out the co-occurrence of wind power density-wind direction (aka. wind energy rose) at the 6 sites (As shown in Fig. 5.11).

Site A: in the upcoming week, the wind power is mainly contributed by the NE and NNE winds while the former obviously takes dominance with occurrences up to around 70% and the later has occurrences around 25%. During the northeasterly wind period, winds with density from 100 W/m^2 to 150 W/m^2 have the highest occurrences (up to 40%), followed by those the density from 300 W/m^2 to 350 W/m^2 (about 20%) and those with density from 350 W/m^2 to 400 W/m^2 (around 10%). Winds with density over 400 W/m^2 are mainly blowing from the northeast.

Table 5.2 Hourly forecasting of wind power density and wind direction at important stations

Month Day	Time	Point A		Point B		Point C		Point D		Point E		Point F	
		WPD (W/m²)	Dir (°)	WPD (W/m²)	Dir (°)	WPD (W/m²)	Dir (°)	WPD (W/m²)	Dir (°)	WPD (W/m²)	Dir (°)	WPD (W/m²)	Dir (°)
07th December	00	94	24	171	45	237	354	449	25	88	56	1200	39
	01	102	23	162	48	226	355	453	26	92	54	1250	41
	02	110	22	153	51	215	356	458	27	95	53	1310	44
	03	119	20	144	54	205	357	463	28	99	51	1370	47
	04	127	19	136	57	195	358	468	29	102	50	1420	50
	05	137	18	128	60	185	359	473	30	106	48	1480	52
	06	146	17	121	63	176	0	478	31	109	47	1550	55
	07	159	16	115	61	178	3	487	31	134	47	1530	55
	08	173	15	110	59	180	5	497	32	161	47	1510	55
	09	187	14	104	57	182	8	507	32	192	48	1490	55
	10	202	13	99	55	184	10	517	32	227	48	1470	54
	11	218	12	94	54	186	13	527	33	266	48	1450	54
	12	235	11	89	52	188	15	537	33	310	49	1430	54
	13	242	12	95	50	177	13	549	34	307	50	1420	55
	14	249	13	100	49	167	11	562	35	304	51	1420	56
	15	257	14	106	47	157	9	575	36	301	52	1420	57
	16	264	15	112	46	147	7	588	37	298	53	1410	58
	17	272	15	119	44	138	5	601	39	295	54	1410	59
	18	280	16	125	43	129	3	614	40	292	55	1410	60
	19	262	16	129	45	129	359	586	41	294	56	1350	58

(continued)

Table 5.2 (continued)

Month Day	Time	Point A WPD (W/m²)	Dir (°)	Point B WPD (W/m²)	Dir (°)	Point C WPD (W/m²)	Dir (°)	Point D WPD (W/m²)	Dir (°)	Point E WPD (W/m²)	Dir (°)	Point F WPD (W/m²)	Dir (°)
	20	246	16	133	46	130	354	558	42	296	57	1300	56
	21	230	16	137	48	130	349	532	43	299	58	1260	53
	22	215	15	141	50	131	345	506	44	301	59	1210	51
	23	200	15	145	52	131	340	481	45	303	60	1160	49
08th December	00	187	15	149	53	132	336	457	46	306	61	1120	47
	01	183	18	144	55	117	334	380	45	257	63	1070	48
	02	179	21	139	57	104	332	313	44	213	65	1020	48
	03	175	24	133	58	92	329	254	43	175	67	976	48
	04	172	27	128	60	80	327	203	42	141	69	932	49
	05	168	30	124	62	70	325	159	41	112	71	889	49
	06	164	33	119	63	61	323	122	40	88	73	847	49
	07	153	35	111	60	51	319	124	38	98	69	859	49
	08	142	37	104	56	43	315	126	36	110	66	870	48
	09	132	39	96	53	36	311	127	33	122	63	882	47
	10	122	41	90	49	29	307	129	31	135	60	894	46
	11	113	42	83	46	24	303	131	28	150	57	906	45
	12	104	44	77	42	19	299	132	26	165	53	918	45
	13	104	48	68	43	18	302	158	22	181	53	920	44
	14	104	53	60	43	18	305	186	18	199	53	922	43

(continued)

Table 5.2 (continued)

Month Day	Time	Point A WPD (W/m²)	Dir (°)	Point B WPD (W/m²)	Dir (°)	Point C WPD (W/m²)	Dir (°)	Point D WPD (W/m²)	Dir (°)	Point E WPD (W/m²)	Dir (°)	Point F WPD (W/m²)	Dir (°)
	15	104	57	53	44	18	308	218	14	218	52	924	42
	16	104	61	46	45	17	311	253	11	238	52	926	41
	17	103	65	40	45	17	314	291	7	259	52	928	40
	18	103	69	35	46	16	317	334	3	281	51	931	39
	19	94	69	37	43	22	316	346	2	271	50	965	37
	20	85	69	39	40	28	314	359	1	260	49	1000	35
	21	76	69	41	37	36	312	371	360	250	47	1040	34
	22	69	69	44	34	45	311	385	359	240	46	1070	32
	23	55	69	49	28	55	309	398	358	230	44	1110	30
…………													
13th December	00	341	41	32	330	414	259	302	208	583	201	754	66
	01	347	40	29	331	438	258	281	207	491	199	726	65
	02	353	40	26	333	463	257	261	205	409	197	699	65
	03	359	39	23	335	489	255	242	203	336	195	673	64
	04	365	39	20	337	516	254	223	201	273	193	647	64
	05	371	39	18	339	543	253	206	199	218	190	622	64
	06	378	38	15	341	572	252	190	197	171	188	598	63
	07	383	38	17	342	585	252	166	196	163	186	589	63
	08	388	38	20	343	598	252	144	194	155	185	581	64

(continued)

Table 5.2 (continued)

Month Day	Time	Point A WPD (W/m^2)	Dir (°)	Point B WPD (W/m^2)	Dir (°)	Point C WPD (W/m^2)	Dir (°)	Point D WPD (W/m^2)	Dir (°)	Point E WPD (W/m^2)	Dir (°)	Point F WPD (W/m^2)	Dir (°)
	09	394	38	22	345	611	252	124	192	147	183	573	64
	10	400	38	25	346	624	251	106	190	139	181	564	64
	11	405	38	27	347	638	251	90	189	132	179	556	64
	12	411	38	30	348	652	251	76	187	125	177	548	64
	13	403	38	32	350	620	251	72	179	157	180	519	63
	14	395	37	34	353	590	252	69	171	194	183	492	62
	15	387	37	36	355	560	252	66	163	236	185	465	61
	16	379	36	38	358	532	252	63	154	284	188	440	60
	17	372	36	40	0	504	252	60	146	338	190	415	59
	18	364	35	42	3	478	252	57	138	398	193	392	58
	19	352	35	49	2	458	251	56	141	376	193	411	58
	20	340	35	58	1	438	250	56	143	354	193	430	58
	21	329	35	67	0	419	249	56	145	333	193	450	58
	22	317	35	77	359	400	247	55	148	313	193	471	58
	23	306	35	89	358	382	246	55	150	294	192	492	58

Fig. 5.11 Forecasting of
wind energy rose
(co-occurrence of wind
power density-wind
direction) for the next week
(start from 00:00 7th,
December 2016)

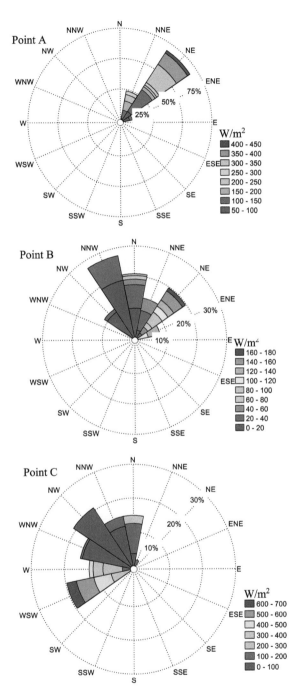

Fig. 5.11 (continued)

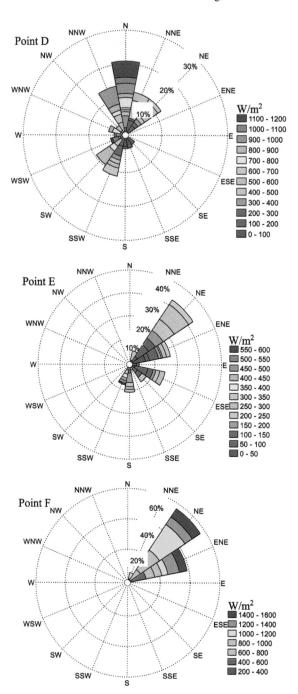

Site B: in the upcoming week, the wind power is mainly contributed by the NE-N-NW winds. It is noteworthy that NNW winds have the highest occurrences (around 26%) while their density is relatively low (basically within 40 W/m^2). The northeasterlies have relatively higher wind power density.

Site C: in the coming week, the wind power is mainly contributed by the WSW-WNW-N winds. The WSW and the NW winds have the highest occurrences (about 26%) while the former's wind power density is higher and the latter lower (basically within 100 W/m^2).

Site D: the NNW-N-NE wind has most of the occurrences, followed by the SW-SSW wind in the following week. The northerlies have the highest occurrences (over 20%). Winds with relatively higher wind power density are mostly from the north and NNW while those with density over 1100 W/m^2 are mainly blowing from the north.

Site E: in the upcoming week, the wind power is mainly contributed by the northeasterlies while the winds with relatively higher wind power density are from the SSW.

Site F: in the following week, almost all the wind power is contributed by the NE and ENE winds.

5.3   Case Study Two

5.3.1   Wind Field Forecast

Based on the wind field forecasts, we mapped out the wind speed and direction in the North Pacific, using the vector arrows to represent wind directions and different background colors to denote wind speed (shown in Fig. 5.12). It is shown that in the upcoming days (from 0:00 on December 26th to 12:00 on December 2011), quite severe tropical cyclone Thane will be exerting its influence on the Bay of Bengal. During that time, the winds in the Bay of Bengal will be spinning anti-clockwise. The cyclone slowly moved towards the northwest and finally landed on the Indian Peninsula. Winds at the middle and southern parts of the Arabian Sea will be steadily blowing from the north to the northeast; those at the Persian Gulf are basically northwesterlies while those at the Gulf of Oman and the tip of the Arabian Sea mostly westerlies. The wind speed of the north of the Arabian Sea will be smaller than that of the middle and the south.

5.3.2   Wind Power Density Forecast

Based on the wind field forecasts and the formula of wind power density, we calculated the wind power density of the end of December in 2011 (shown in Fig. 5.13). In

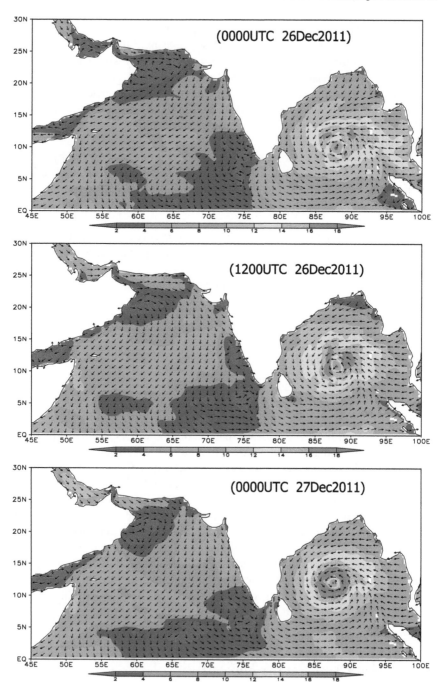

Fig. 5.12 Forecasting of wind field for the next 5 days (start from 00:00 26th, December 2011) of the North Indian Ocean, unit of wind speed: m/s

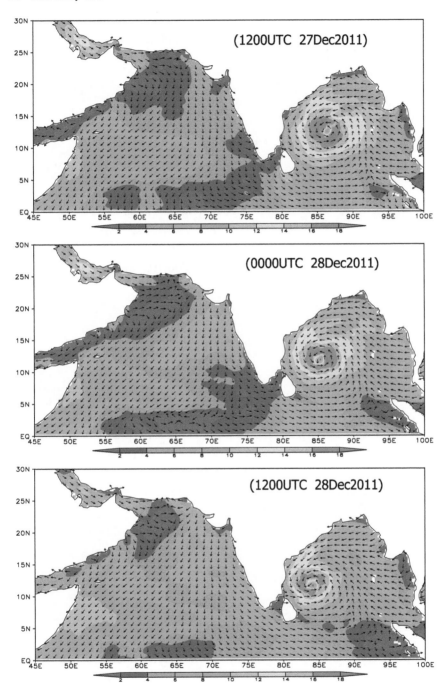

Fig. 5.12   (continued)

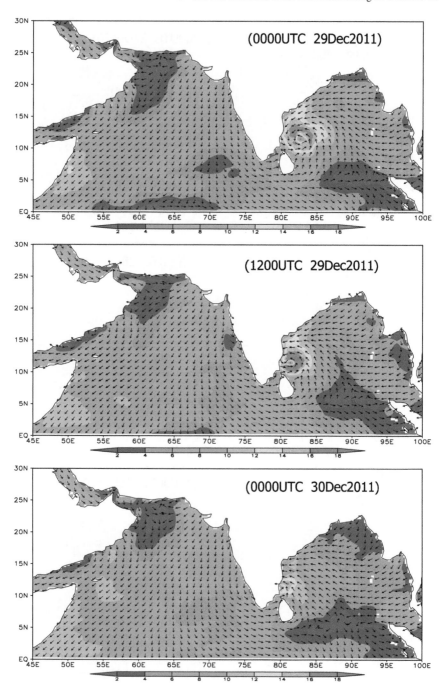

Fig. 5.12 (continued)

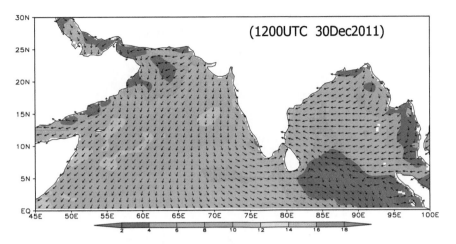

Fig. 5.12 (continued)

the Bay of Bengal, under the influence of the tropical cyclone, most of the sea areas
have the wind power density over 600 W/m^2 while the lowest values are found at the
pinhole. In the next five days, the wind power density of the Arabian Sea is obviously
lower than that of the Bay of Bengal while relatively higher winds are found around
Somali with their density up to 600 W/m^2 under the impact of the cold air.

5.3.3 Availability Forecast

Based on the wind field forecasts from 0:00 on December 26th to 12:00 on December
30th, we calculated the EWSO in the North Indian Ocean, so as to predict the availability of the wind power during the same time period (as shown in Fig. 5.14). In
the Bay of Bengal, in the upcoming five days the availability is quite promising
which is basically over 50%; sea areas with availability over 90% would take up
more than half of the Bay and those with relatively lower availability are distributed
at the eastern and northern coastal areas. In the Arabian Sea, the availability of the
northern part is less than 20% which is way smaller than that of the middle and the
south of the Sea (which is above 50% and even up to 90% at the center).

5.3.4 Energy Level Occurrence Forecast

Based on the wind power density from 0:00 on December 26th to 12:00 on December
30th, we calculated the energy level occurrences in the North Indian Ocean during
Thane (as shown in Fig. 5.15).

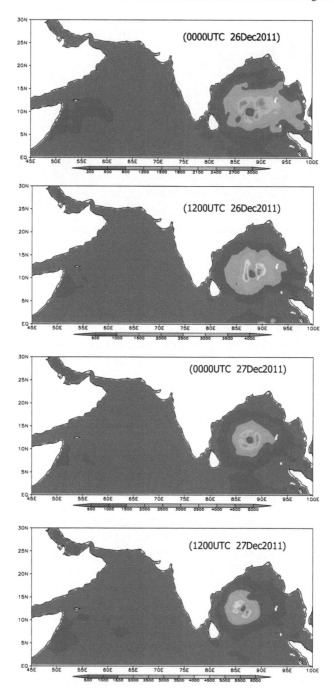

Fig. 5.13 Forecasting of wind power density for the next 5 days (start from 00:00 26th, December 2011) of the North Indian Ocean, unit: W/m^2

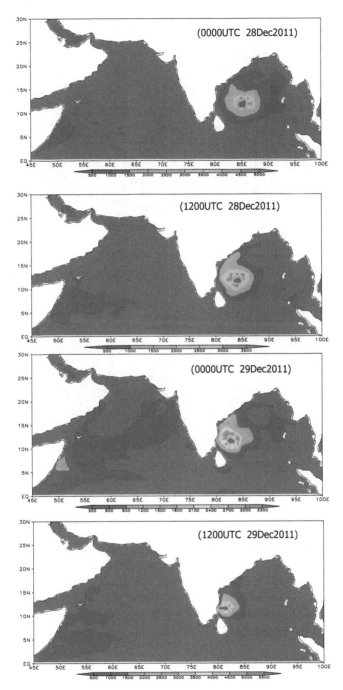

Fig. 5.13 (continued)

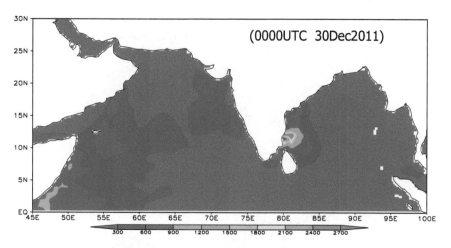

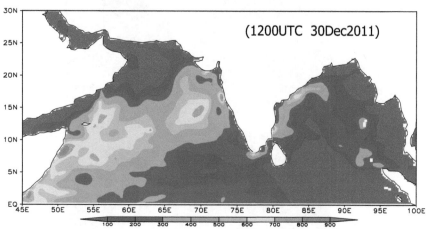

Fig. 5.13 (continued)

ALO: for most of the sea areas in the Bay of Bengal, the occurrences are above 50% with values exceeding 90% at a wide range. The occurrences at the inlet of the Arabian Sea are within 10%; those at the middle and the southern parts are over 40% and those of most sea areas over 90%. The spatial distribution of the MLO, RLO, ELO and SLO is mostly in line with that of ALO only with smaller values. Overall, the ELO of the Bay of Bengal and the surrounding sea areas of Somali is quite promising in the upcoming five days.

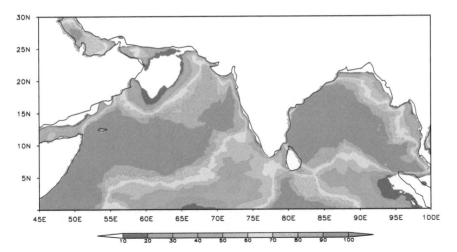

Fig. 5.14 Forecasting of wind energy availability (effective wind speed occurrence) for the next 5 days (start from 00:00 26th, December 2011) of the North Indian Ocean, unit: %

5.3.5 Energy Storage Forecast

In reference to the methods in Sect. 5.2.5, we calculated the total storage, exploitable storage and technological storage of wind energy in the North Indian Ocean during Thane (as shown in Fig. 5.16). In the following five days, the storage of the Bay of Bengal is obviously higher than that of the Arabian Sea.

Total storage: in the Bay of Bengal, the contour lines form a circle with the total amount exceeding 4×10^4 W·h/m^2 in most sea areas and 18×10^4 W·h/m^2 at the center. Lower values are found at the inlet of the Bay and eastern coastal areas. In the Arabian Sea, the large values are found surrounding Somali (6×10^4 W·h/m^2 and above) with contour lines running in a northeast-southwest direction.

Exploitable storage and the technological storage: their spatial distribution is generally in line with that of the total storage only with smaller values.

5.3.6 Forecast of Key Point

Sri Lanka sits in the middle of the North Indian Ocean's main channel. Hereby we randomly assume there is a site around Sri Lanka and carry out wind power forecasts on wind power including hourly wind power density, wind directions and wind energy rose (as shown in Fig. 5.17).

(1) **Hourly WPD and wind direction**

Hourly wind power density and wind directions are shown in Fig. 5.18 and Table 5.3. During Thane from 0:00 on December 26th to 0:00 on 28th, the wind power

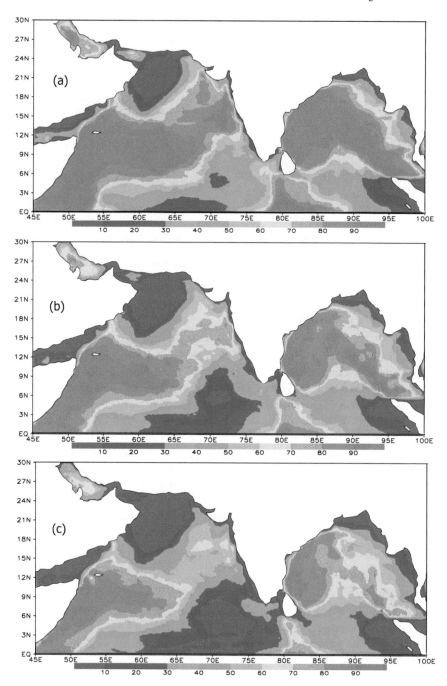

Fig. 5.15 Forecasting of energy level occurrences with wind power density above 100 W/m^2 **a**, 150 W/m^2 **b**, 200 W/m^2 **c**, 300 W/m^2 **d**, 400 W/m^2 **e** for the next 5 days (start from 00:00 26th, December 2011) of the North Indian Ocean, unit: %

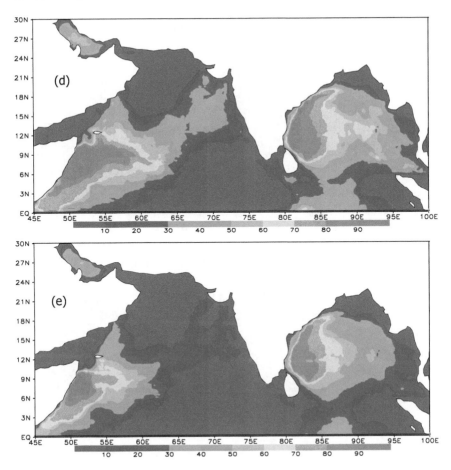

Fig. 5.15 (continued)

density at the site maintains above 200 W/m^2 steadily but presents a slow decreasing trend. The wind power density stays stable from 0:00 on 28th to 21:00 on 30th. From 0:00 on 26th to 0:00 on 28th the wind comes from 300 degrees and then the wind direction changes from north to southwest and then to south. Detailed data is shown in Table 5.3.

(2) **Wind energy rose**

Based on the hourly wind power density and wind directions, we mapped out the WPD-direction joint occurrences, aka. wind energy rose (as shown in Fig. 5.19). In the next five days, the wind power is mainly contributed by winds blowing from the EW and the WNW, followed by those from the SSW. Winds with density over 800 W/m^2 basically come from the northwest. And as for northwesterlies, those with density from 300 W/m^2 to 400 W/m^2 have the highest occurrences, followed by those with density from 600 W/m^2 to 700 W/m^2. For winds coming from the WNW, the

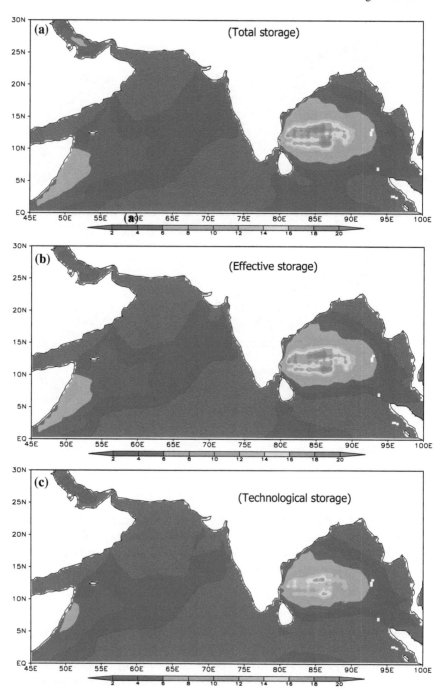

Fig. 5.16 Forecasting of total storage **a**, exploitable storage **b**, technological storage **c** for the next 5 days (start from 00:00 26th, December 2011) of the North Indian Ocean, unit: %

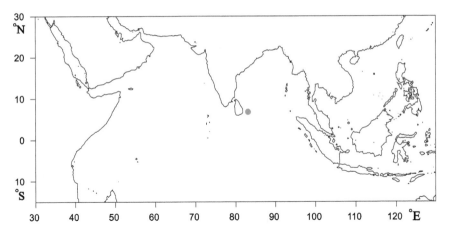

Fig. 5.17 Distribution of site

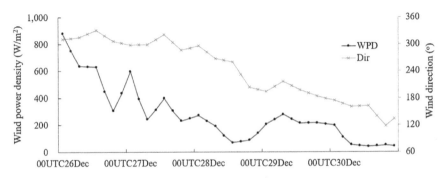

Fig. 5.18 Forecasting of 3-hourly wind power density and wind direction at one site (start from 00:00 26th, December 2011) of the North Indian Ocean

most frequent winds are those with density from 200 W/m^2 to 300 W/m^2 while for those coming from the SSW are 200 W/m^2 to 300 W/m^2.

5.4 Summary

In this Chapter, an all-elements short-term forecasting scheme for wind energy was proposed, with the Maritime Silk Road as a case study. The main results are:

(1) We could transform from simple weather forecasting to the combination of weather and wind power forecasting, so as to carry out intensive wind power forecasting.

Table 5.3 Hourly forecasting of wind power density and wind direction at one station in the surrounding waters of Sri Lankan

Time	WPD (W/m²)	Dir (°)	Time	WPD (W/m²)	Dir (°)	Time	WPD (W/m²)	Dir (°)
00:00UTC 26Dec	880.51	311.425	00:00UTC 28Dec	271.76	296.5	00:00UTC 30Dec	198.52	174.951
03:00UTC 26Dec	752.2	313.334	03:00UTC 28Dec	230.05	282.289	03:00UTC 30Dec	110.14	167.736
06:00UTC 26Dec	637	315.243	06:00UTC 28Dec	192.84	268.078	06:00UTC 30Dec	52.87	160.521
09:00UTC 26Dec	633.83	323.36	09:00UTC 28Dec	120.71	264.306	09:00UTC 30Dec	46.11	161.489
12:00UTC 26Dec	630.67	331.477	12:00UTC 28Dec	69.27	260.535	12:00UTC 30Dec	39.95	162.458
15:00UTC 26Dec	449.95	319.396	15:00UTC 28Dec	78.41	232.399	15:00UTC 30Dec	44.76	140.559
18:00UTC 26Dec	307.54	307.314	18:00UTC 28Dec	88.32	204.263	18:00UTC 30Dec	49.96	118.661
21:00UTC 26Dec	437.07	302.711	21:00UTC 28Dec	139.42	199.225	21:00UTC 30Dec	44.07	134.018
00:00UTC 27Dec	598.65	298.107	00:00UTC 29Dec	207.18	194.186			
03:00UTC 27Dec	394.49	298.733	03:00UTC 29Dec	241.51	205.751			
06:00UTC 27Dec	242.98	299.359	06:00UTC 29Dec	279.43	217.315			
09:00UTC 27Dec	314.55	310.287	09:00UTC 29Dec	245.7	207.516			
12:00UTC 27Dec	398.95	321.215	12:00UTC 29Dec	214.81	197.717			
15:00UTC 27Dec	307.6	304.076	15:00UTC 29Dec	215.82	190.867			
18:00UTC 27Dec	231.37	286.937	18:00UTC 29Dec	216.85	184.017			
21:00UTC 27Dec	251.03	291.718	21:00UTC 29Dec	207.55	179.484			

(2)   In terms of parameters, besides the wind power density and wind field forecasts which have already been given attention to, it is necessary to carry out short-term forecasts on key parameters such as availability, energy level occurrences and storage amount. This chapter has carried out the short-term wind power forecasts based on two cases of tropical cyclone influencing the Bay of Bengal (as well as the case of cold air influencing the South China Sea). And forecasts systematically covered key parameters such as hourly forecasts on surface wind field and wind power density, the availability over the upcoming week (aka the

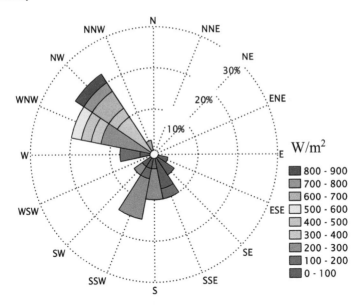

Fig. 5.19 Forecasting of wind energy rose (co-occurrence of wind power density-wind direction) for the next 5 days (start from 00:00 26th, December 2011)

occurrences of winds at a speed from 5 to 25 m/s next week), energy level occurrences (the occurrences of winds with density of 100, 150, 200, 300 and 400 and above next week), energy storage (total storage, the exploitable storage and the technological storage), and we carried out wind power forecasting at key sites including hourly wind power density, wind directions and wind energy roses.

(3)   In the South China Sea, at the end of autumn and early spring, the constant, strong and stable cold air has brought abundant and stable wind power to the South China Sea, especially to the middle and southern parts. In the North Indian Ocean, the tropical cyclone will bring winds with relatively higher wind power density and the EWSO is generally high during the tropical cyclone and the cold air period.

(4)   In this chapter, we conducted short-term wind power forecasts over a period of five to seven days. This forecast scheme can be adjusted to difference time span such as 10 days, two weeks or half a month days according to practical needs.

This Chapter also found an important phenomenon of tropical cyclone. Based on the movement directions of the cyclone, when it is at an open sea area, fast winds are mainly located at the right side (the first and fourth quadrant); otherwise, strong winds are generally found at the front (the first and second quadrants) when the moving direction of cyclone was blocked by the terrain.

References

Bailey B, Brower MC, Zack J (1999) Short-term wind forecasting (ISBN 1902916X). In: Proceedings of the European wind energy conference, Nice, Frace, Mar 1–5, 1999, pp 1062–1065

EPRI (2003) California wind energy forecasting system development and testing, phase 1: initial testing. EPRI Final Report 1007338, Jan 2003

Focken U, Lange M, Waldl HP (2001) Previento—a wind power prediction system with an innovative upscaling algorithm. In: Proceedings of the European wind energy conference, Copenhagen, Denmark, June 2–6, 2001, pp 826–829

Kariniotakis G, Moassafir J, Usaola J (2003) ANEMOS: development of a next generation wind resource forecasting system for the large-scale integration of onshore & offshore wind farms. In: European wind energy conference & exhibition EWEC 2003. Madrid, Spain

Lars L, Watson SJ (1994) Short-term prediction of local wind conditions. Bound-Layer Meteorol 70:171

Lars L (1999) Short-term prediction ofthe power production from wind farms. J Wind Eng Ind Aerodyn 90:207–220

Milligan M, Schwartz M, Wan Y (2003) Statistical wind power forecasting for U.S. wind farms. Preprint of the conference "WINDPOWER 2003", Austin, May 18–21 2003

Nielsen TS, Madsen H, Nielsen HA (2006)Short-tem wind power forecasting using advanced statistical models. Eur Wind Energy Conf

Pinard JP, Benoit R, Yu WA (2005) West wind climate simulation of the mountainous Yukon. Atmos Ocean 43(3):259–282

Yu W, Benoit R, Girard C (2006) Wind Energy Simulation Toolkit (WEST): a wind mapping system for use by the wind-energy industry. Wind Eng 30(1):15–33

Zheng CW, Xu JJ, Zhan C, Wang Q (2019) In: 21st Century maritime silk road: wave energy resource evaluation. Springer

Zheng CW, Zhou L, Song S, Su Q (2014) Forecasting of the China Sea wind energy density. J Guangdong Ocean Univ 34(1):71–77

Zheng CW, Li CY, Pan J et al (2016) An overview of global ocean wind energy resources evaluation. Renew Sustain Energy Rev 53:1240–1251

Chapter 6
An All-Elements Long-Term Projection of Offshore Wind Energy in the Maritime Silk Road

As early as the end of the last century, some developed countries began to pay attention to offshore wind energy. Despite the extremely scarce ocean data, they still made great contributions. This stage is mainly based on extremely limited site observation data, ship report data, etc., to carry out offshore wind energy assessment for single stations and small coastal areas. Benefiting from the rapid development of ocean observation methods and computer technology, more and more satellite data, simulation data, and reanalysis data are widely used in offshore wind energy evaluation. Nowadays, humans can make comprehensive and three-dimensional statistics on the climate characteristics of wind energy resources in the global seas, and analyze the historical trend of wind energy, resource reserves, resource level zoning, etc. (Ma et al. 2021; Li et al. 2020; Zheng et al. 2013; Zheng and Pan 2014), and use numerical models to develop short-term forecasts of wind energy (Tambke et al. 2005). The analysis of climatic characteristics of wind energy can provide a scientific basis for the site selection of wind farms, and short-term forecasts can provide references for the operational operation of wind turbines. However, so far the research on medium and long-term forecasting of wind energy is still extremely scarce, although this work is closely related to the medium and long-term planning of wind energy development and power dispatching.

Zheng et al. (2019) proposed an intensive wind energy prediction method: applying the medium and long-term forecasting wind field data interpretation to wind energy forecasting, mainly using the CMIP5 wind field to predict the wind energy from 2080 to 2099, and comparing it with the historical status between 1980 and 1999, to provide a scientific basis for the medium and long-term planning of wind energy development. Zheng et al. (2017) used artificial neural networks and linear continuation to carry out the medium and long term estimate of wind energy resources in the Gwadar Port, which is on the basis of mastering the climatic variation characteristics of wind energy. The above work provides a technical approach for the mid- and long-term estimation of wind energy resources. Zheng et al. (2016) pointed out that the following three types of methods can be used for medium and

C. Zheng et al., *21st Century Maritime Silk Road: Wind Energy Resource Evaluation*, Springer Oceanography, https://doi.org/10.1007/978-981-16-4111-4_6

long-term projection of wind energy in the future: (1) Medium- and long-term estimation of wind energy based on the relationship between wind energy and important factors. (2) using of least squares support vector machine, artificial neural network, Hilbert to carry out the wind energy projection. (3) using CMIP data to make medium and long-term projection of wind energy. This chapter uses the CMIP5 wind field data to develop a mid- and long-term forecast of the wind energy resources of the Maritime Silk Road for the next 40 years, to provides a scientific basis and auxiliary decision-making for the mid- and long-term planning and optimization of wind energy development.

6.1 Data and Methods

6.1.1 Data

CMIP5 wind field data: This data comes from the IPCC, and the future global wind field is numerically simulated by multiple countries and multiple models. CMIP5 data is widely used in the future climate projection (Taylor et al. 2012; Xie et al. 2013; Huntingford et al. 2013; Wang et al. 2014). In addition, CMIP5 data is also widely used in the research of ecosystems, flood, tropical storms, monsoon, etc. (Villarini and Vecchi 2012; Poulter et al. 2014; Hirabayashi et al. 2013; Ramesh 2014). De Winter et al. (2013) used 12 experimental data with good results to analyze the extreme wind speed of the North Sea Basin. Similarly, this article obtains the global sea surface wind field for the next decades and performs the wind energy projection of the Maritime Silk Road. The data is processed into the following format: the spatial range is: 89°S ~ 89°N, 1.25° ~ 358.75°E, the spatial resolution is 2.5° × 2.0°, and the time series is 00:00 on January 1, 2015 to 21:00 on December 31, 2100, the time resolution is 3-hourly (one data every three hours).

ERA-Interim reanalysis data: the introduction is shown in Chap. 3.

6.1.2 Method

This chapter aims to design an all-elements mid-long term projection scheme for wind energy resource, to provide reference for the long term planning of wind energy utilization. At present, there are abundant researches on the medium and long-term projection of meteorological and oceanic elements. However, the research on the projection of offshore wind energy is scarce, and the research focusing on the Maritime Silk Road is even rarer, and it cannot provide a good scientific reference for the medium and long-term wind energy development plans. In addition, the traditional long-term projection of wind energy mainly focused on the wind power density and wind speed. In this Chapter, we emphasis that: Whether wind energy climate characteristics analysis or, wind energy short-term forecasting or long-term wind energy projection, it is necessary to pay attention to a set of key indicators of wind energy, systematically including a set of key indicators of wind energy (wind power density, energy availability, energy level occurrences, energy stability, energy storage, etc.).

We firstly carry out the long-term wind energy projection for the next decades. We also analyze the wind energy characteristics for the past decades. Then the wind energy characteristics for the next decades and the past decades are compared, to provide scientific reference for the long-term plan of wind energy utilization. Using the 3-hourly CMIP5 wind field data from 2020 to 2059, the 3-hourly wind power density for the next 40 years is calculated. Then the wind energy resources of the Maritime Silk Road for the next 40 years was projected, systematically covering the wind power density, energy availability, energy level occurrences, energy stability, energy storage, etc. At the same time, using the ERA-Interim reanalysis data to calculate the wind energy characteristics of the historical status of this sea area, and comparing the wind energy indicators of the future and the historical status, to provide a reference for the medium and long-term planning of wind energy resource development.

6.2 Seasonal Characteristics of Wind Energy Density

The wind power density is the most direct manifestation of wind energy resources. Here we use the 3-hourly wind power density from 2020 to 2059 to calculate and analyze the seasonal characteristics of wind energy in the Maritime Silk Road for the next 40 years. At the same time, using the 6-hourly WPD from 1979 to 2015, to calculate and analyze the seasonal characteristics of the wind power density of the Maritime Silk Road for the past near 40 years (representing the historical status). Then the wind energy of the next 40 years and that of the past near 40 years are compared and analyzed, as shown in Fig. 6.1.

The seasonal difference of wind power density in the Maritime Silk Road for the next 40 years is reflected in the following: In the Arabian Sea, the wind power density is the largest in August, followed by February and November, and the lowest in May. In the Bay of Bengal, August is the highest, followed by May and November, and February is the lowest. In the South China Sea, WPD in February and November was significantly higher than that in May and August.

In February (representing winter, the same below): Regardless of the South China Sea, the Bay of Bengal, or the Arabian Sea, the future wind power density will be significantly higher than that in the past, which means that there will be signs of overall enhancement of the cold air in the Maritime Silk Road in the future. In this month, the wind power density in the South China Sea is the largest, followed by the Arabian Sea, and the Bay of Bengal is relatively the smallest, both in the past and in the future. In the South China Sea, the average wind power density in most sea areas for the next 40 years will be above 300 W/m^2, which is significantly higher

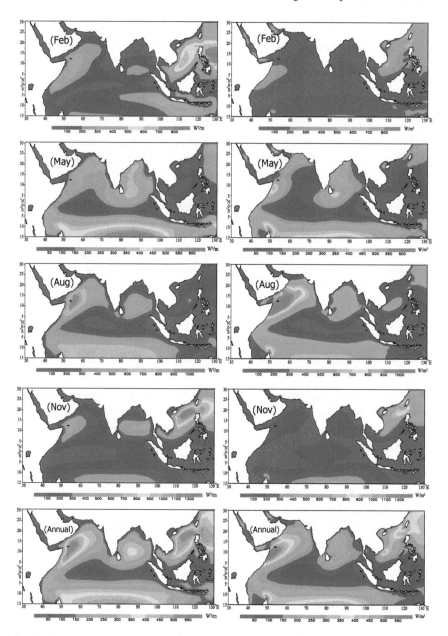

Fig. 6.1 Wind power density for the next 40 years (left) and the past near 40 years (right)

than the historical status (above 200 W/m^2). In the Arabian Sea, large-value areas are distributed in the coastal waters of Somalia, and the average wind power density in the next 40 years will be above 200 W/m^2, which is significantly higher than the historical status (above 100 W/m^2). In the Bay of Bengal, the average wind power density for the next 40 years is basically within 300 W/m^2, and the historical wind power density is basically within 200 W/m^2.

In May (representing spring, the same below): the wind power density in the Bay of Bengal is the highest, followed by the Arabian Sea, and the South China Sea is relatively the lowest, both in the past and in the future. Zheng (2018) pointed out that the South China Sea was in the transition season from the northeast monsoon to the southwest monsoon in this month, but the Arabian Sea and the Bay of Bengal had completed the transition from the northeast monsoon to the southwest monsoon. In the South China Sea, the historical status is close to the future status, and the wind power density in the central and northern regions is 50–150 W/m^2. In the Bay of Bengal, the average wind power density for the next 40 years in a large area is above 200 W/m^2, which is significantly higher than the historical status (above 150 W/m^2), especially the area range with wind power density above 350 W/m^2 is obviously wider than the historical status. However, the wind power density of the historical status has a value of more than 400 W/m^2, but it does not exist in the future, which means that the occurrence of the southwest monsoon surge in the Bay of Bengal in the coming season will be higher than in the historical status, but the intensity will be weaker. In the Arabian Sea, the wind power density in the historical status has a range of more than 400 W/m^2, but the wind power density in the next 40 years does not exist. The area range of wind power density above 150 W/m^2 for the next 40 years basically covers the entire Arabian Sea, which is obviously wider than the historical status. This means that the occurrence of the southwest monsoon surge in the Arabian Sea in this season will be higher than in the past, but the intensity will be weaker.

In August (representing summer, the same below): Regardless of the South China Sea or the North Indian Ocean, the average wind power density for the next 40 years is significantly smaller than the historical status. In the South China Sea, the high-value areas are distributed in the traditional gale center of the South China Sea (southeast of the Indo-China Peninsula). The historical wind power density is above 300 W/m^2, but the average wind power density for the next 40 years is significantly lower (above 200 W/m^2). In the Bay of Bengal, historical wind power density above 200 W/m^2 basically cover the entire Bay of Bengal, but the future wind power density ranges above 200 W/m^2 will be significantly smaller. In the Arabian Sea, the future wind power density range of 900 W/m^2 is significantly smaller than the historical status.

In November (representing autumn, the same below): The South China Sea and the North Indian Ocean have basically completed the transition from the southwest monsoon to the northeast monsoon. Regardless of the South China Sea, the Bay of Bengal or the Arabian Sea, the average wind power density for the next 40 years is higher than the historical status, which means that the cold air along the Maritime Silk Road will increase in the future.

Annual average: The multi-year average wind power density of the South China Sea and the Bay of Bengal for the next 40 years is significantly higher than the historical status, and the multi-year average wind power density in the Arabian Sea for the next 40 years is close to the historical status. In the South China Sea, the average wind power density in the central and northern regions for the next 40 years will basically be above 350 W/m², the Gulf of Thailand will basically be above 150 W/m², and the Beibu Gulf will be the low-value center. In the Bay of Bengal, the average wind power density for the next 40 years will basically be above 150 W/m², and the center will reach above 350 W/m². In the Arabian Sea, most areas are above 200 W/m².

6.3 Availability of Wind Energy

It is generally believed that the wind speed between 5 and 25 m/s is beneficial to the collection and conversion of wind energy resources, and the wind speed in this interval is defined as the effective wind speed for wind energy development (abbreviated as effective wind speed). Obviously, the effective wind speed occurrence (EWSO) directly reflects the availability of wind energy. Here we use the 3-hourly wind speed data from 2020 to 2059 to statistically analyze the seasonal characteristics of the EWSO of the Maritime Silk Road for the next 40 years. At the same time, we use the 6-hourly wind speed data from 1979 to 2014 to do statistic analysis EWSO of the Maritime Silk Road for the past decades. The seasonal characteristics of EWSO in the past near 40 years (representing the historical status of EWSO). Then a comparative analysis of EWSO for the next 40 years and the past near 40 years is shown in Fig. 6.2.

The seasonal difference of the EWSO of the Maritime Silk Road for the next 40 years is reflected in the following: In the Arabian Sea, the EWSO is the highest in August, followed by February and November, and the lowest in May. In the Bay of Bengal, August and November are the highest, followed by May, and February is the lowest. In the South China Sea, February is the highest, November is the second, and May is the lowest. The detail information is as follows,

In February, the resource utilization rate in the South China Sea is the highest, followed by the Arabian Sea, and the Bay of Bengal is relatively the lowest. The values in the past decades and the future decades are almost unchanged. In the South China Sea, the average EWSO for the next 40 years in a large area will be above 95%, which is significantly higher than the historical status (70%-85%). In the Bay of Bengal, the EWSO in the top region is relatively low, basically within 30%. And the central and southern regions of the Bay of Bengal are relatively high, with the EWSO in the next 40 years 40–85%. In the Arabian Sea, the EWSO in most regions in the next 40 years will be more than 70%, the center will reach more than 95%, and the area range of more than 95% is also significantly higher than the historical status.

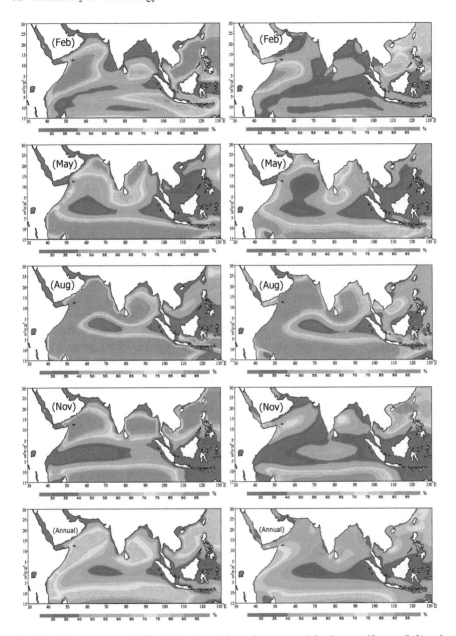

Fig. 6.2 Wind energy availability (effective wind speed occurrence) for the next 40 years (left) and the past near 40 years (right)

In May, the EWSO in the South China Sea for the next 40 years is lower than its historical status, while the situation in the Bay of Bengal and the Arabian Sea is the opposite. In the South China Sea, the transition from the northeast monsoon to the southwest monsoon has not yet been completed. The northern part of the South China Sea is still significantly affected by the cold airs, resulting in a significantly higher EWSO in the northern part of the South China Sea than in the central and southern parts. In the Bay of Bengal, the EWSO in most areas will be above 55% for the next 40 years, with the center located in the eastern offshore of the Indian Peninsula (above 90%), followed by southern Sri Lanka (85%-90%). In the Arabian Sea, the high-value areas of EWSO for the next 40 years will be distributed in the northeastern coastal waters (over 90%) of the Arabian Sea, while the historical high-value areas will be distributed in the northern and northwestern coastal waters of the Arabian Sea.

In August, under the influence of the strong southwest monsoon, the EWSO of almost the entire Arabian Sea is above 95%, both in the future and in the past. The average EWSO in most of the Bay of Bengal for the next 40 years will be above 70%, from southwest to northeast gradually decreasing. The overall EWSO in the South China Sea for the next 40 years is lower than the historical status.

In November, under the influence of cold airs, the EWSO in the central and northern parts of the South China Sea is significantly higher than that in the southern part. This phenomenon is almost the same in the future and in the past. In the next 40 years, the EWSO in the central and northern parts of the South China Sea will exceed 95%. The EWSO in the Bay of Bengal and the Arabian Sea for the next 40 years is also significantly higher than the historical status, with a large area of over 95%, while the historical state is basically below 80%.

Annual average: The EWSO of the Maritime Silk Road for the next 40 years will be more than 65% (more than 75% in the center), which is higher than the historical status. In the South China Sea, the future EWSO in most of the central and northern regions is above 70%, the Beibu Gulf is a low-value center, and the future EWSO in the Taiwan Strait is also significantly lower than the historical status. In the Bay of Bengal, the future EWSO in most areas is above 65%, and the contour is roughly circular. The historical EWSO contour is southwest-northeast, and the value is significantly lower than the average value of the next 40 years. In the Arabian Sea, the average EWSO for the next 40 years is basically above 65%, which is significantly higher than the historical status.

6.4   Energy Level Occurrences

It is generally considered that wind power density above 200 W/m^2 is rich in resources. Obviously, the energy level occurrence of wind power density above 200 W/m^2 (abbreviated as rich level occurrence, RLO) reflects the abundance of resources. Here we use the 3-hourly wind power density data from 2020 to 2059 to statistically analyze the seasonal characteristics of RLO in the Maritime Silk Road for the next 40 years; At the same time, using the 6-hourly wind power density data from 1979 to 2015 to statistically analyze the seasonal characteristics of the RLO in the Maritime Silk Road for the past near 40 years (representing the past climate characteristics of the RLO). And then the RLO for the next 40 years and the past near 40 years are compared and analyzed, as shown in Fig. 6.3.

The RLO of the Maritime Silk Road for the next 40 years shows a large seasonal and regional difference: in the Arabian Sea, the RLO is the highest in August, followed by November, and the lowest in May. In the Bay of Bengal, August is the highest, November is the second, and February is the lowest. In the South China Sea, the RLO was the highest in February, followed by November, and the lowest in May.

In February: Regardless of the South China Sea or the North Indian Ocean, the average RLO for the next 40 years is significantly higher than the historical status. In the South China Sea, the overall RLO is optimistic. In the next 40 years, most areas will be above 80%, and the center will be located in the southeastern waters of the Indochina Peninsula (up to 90%). In the Bay of Bengal, the high-value areas are distributed in the central and southern regions (the average status in the next 40 years is 20%-50%, and the historical status is 10%-30%). In the Arabian Sea, the average RLO for most regions for the next 40 years will be above 20% (The historical state is that most regions are above 10%). The center is located in the waters near Somalia (above 90%).

In May: In the South China Sea, the future RLO is basically within 30%, and there are 3 relatively large value areas, the Luzon Strait and its western part, the traditional gale center of the South China Sea, and the Gulf of Thailand. In the Bay of Bengal, the future RLO of most regions is above 30%, and the center reaches 80%. In the Arabian Sea, the future RLO of most regions is above 20%, and the center is located in the coastal area of Somalia-the top of the Arabian Sea.

In August: The overall RLO in the Arabian Sea and the Bay of Bengal is optimistic, and the area range of above 80% for the next 40 years is obviously wider than the historical status. The future RLO of the South China Sea is slightly lower than the historical status.

In November: The future RLO in the Maritime Silk Road is more optimistic than the historical status. The RLO in the central and northern parts of the South China Sea is as high as 90%, and the southern part is basically within 70%. In the Bay of Bengal, the future RLO of the central and southern regions is above 40%, the center is above 80%, and the contours are distributed in an east–west band. In the Arabian Sea, the future RLO (above 30%, the center is 90%) is significantly higher than the historical status (above 10%, the center is more than 40%).

Annual mean: The RLO of the Maritime Silk Road for the next 40 years is generally higher than the historical status. In the South China Sea, the future RLO of most regions is more than 30%, and the RLO of more than 50% is wider than the historical status. In the Bay of Bengal, the RLO of most areas is above 20%, and more than 50% of the sea area in the future will be wider than the historical status. In the Arabian Sea, the future RLO of most sea areas is above 40%, which is about 10% higher than

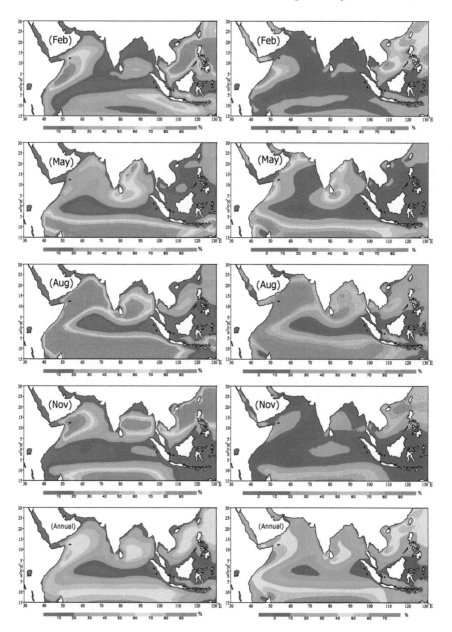

Fig. 6.3 Energy level occurrences of wind power density above 200 W/m^2 for the next 40 years (left) and the past near 40 years (right)

the historical state, and the range of large value areas above 60% is also wider than the historical state.

6.5 Energy Stability

Based on the 3-hourly wind power density data in February 2020 and calculation method of coefficient of variation (Cv), the Cv in February 2020 is calculated. Similarly, the Cv in every February for the next 40 years is calculated. Then the multi-year average Cv in February is obtained. Using the same method, the multi-year average Cv in February, May, August and November for both the past near 40 years and the next 40 years are obtained, as shown in Fig. 6.4. Then we compare and analyze the past and future wind energy stability of the Maritime Silk Road. The smaller the Cv, the better the stability of wind energy, and vice versa, the worse the stability.

On the whole, the spatial distribution of the future Cv and the past Cv in each month remains roughly the same, but the annual average Cv for the next 40 years is significantly smaller than the historical status, indicating that the wind energy stability of the next 40 years is better than that in the past near 40 years as a whole, which is beneficial to the resource development. The resource stability of the Maritime Silk Road for the next 40 years shows large seasonal differences: in the Arabian Sea, August has the best stability, followed by February and November, and May has the worst stability. In the Bay of Bengal, August has the best stability, followed by November, and February has the worst stability. In the South China Sea, February has the best stability, followed by November, and May has the worst stability.

In February: The Cv in the South China Sea for the next 40 years is basically within 0.8, which is significantly lower than the historical status (above 0.9). The Cv in the Bay of Bengal for the next 40 years is basically within 1.1, which is significantly lower than the Cv of the historical status (the center can reach 1.3). The Cv in most parts of the Arabian Sea for the next 40 years is within 0.8, which is significantly lower than the historical status. As a whole, the energy stability in the future February is better than that in the past February in the Maritime Silk Road.

In May: In the South China Sea, the historical Cv of most regions is above 1.0. The large value area with Cv above 1.3 is wide, basically covering the central and eastern South China Sea. And the future Cv will be significantly lower (most areas are within 1.1). The Cv of the historical state of the Bay of Bengal and the Arabian Sea is significantly higher than the average for the next 40 years by about 0.4.

In August: In the South China Sea, the historical large Cv areas are distributed in the northern part of the South China Sea (above 1.3), which is significantly higher than the values of the next 40 years (within 0.9). In the Bay of Bengal, the stability of the top region is worse than that of the central and southern parts. In the Arabian Sea, the projection Cv for the next 40 years in a wide range of waters is within 0.4, while in the historical status, only the Cv in the offshore Somalia is within 0.4. The sea area near the equator of the North Indian Ocean is a large value area of Cv, with a west–east belt spatial distribution.

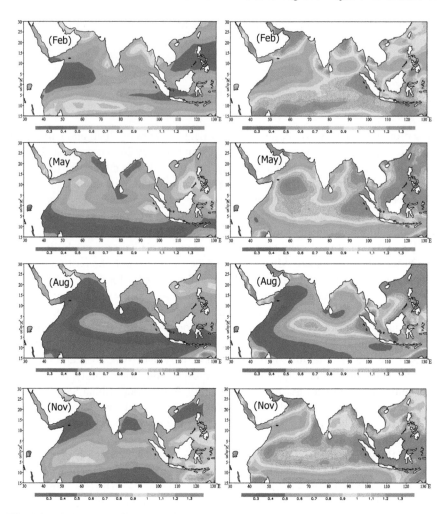

Fig. 6.4 Wind energy stability (coefficient of variation) for the next 40 years (left) and the past near 40 years (right)

In November: In the South China Sea, the low value center of Cv is mainly located from the Luzon Strait to the Hainan Island, both in the past and in the future. The projection Cv along the Maritime Silk Road (below 0.7) is obviously smaller than that of the historical status.

6.6 Monthly/seasonal Variation of Energy

Refering to the algorithm of Cornett (2008) to calculate the Mv year by year from 2020 to 2059, and then getting the multi-year average Mv, as shown in Fig. 6.5a. In the same way, the Mv of the historical status is obtained, as shown in Fig. 6.5b. The greater the value of the monthly variation index, the more significant of the monthly difference in wind energy, which is not conducive to wind energy development. On the contrary, the smaller the monthly difference in wind energy. The Mv of the next 40 years and the past near 40 years is compared to show the change of the monthly difference of wind energy.

Mv: As shown in Fig. 6.5, the spatial distribution of the future Mv and the past Mv are roughly consistent as a whole. The Mv of the Arabian Sea is significantly larger

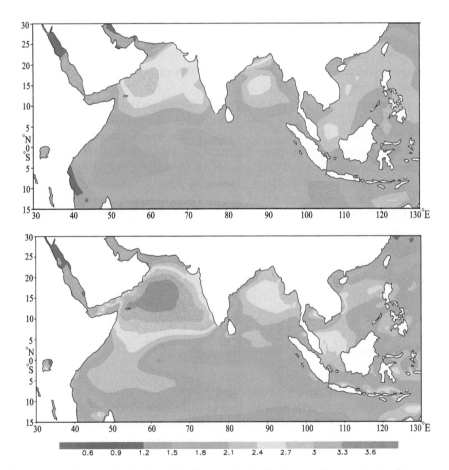

Fig. 6.5 Monthly variation index of wind power density for the next 40 years (top) and the past near 40 years (bottom)

than that of the Bay of Bengal and the South China Sea. In the South China Sea, the projection Mv is roughly the same as the Mv value of the historical status, but the area range of projection Mv above 2.1 is wider than that of the historical status, which means that the future monthly difference of wind energy in this sea area will increase. In the Bay of Bengal, the projection Mv is significantly smaller than the historical state, and the large value area above 2.4 is significantly smaller than the historical state. It means that the future monthly difference of wind energy in this sea area will decrease, which is benefit for wind energy utilization. In the Arabian Sea, the projection Mv is basically within 3.0, but the historical Mv of the sea area is basically above 2.4 (with large value center above 3.3). The projection Mv in the Bay of Bengal and the Arabian Sea tends to decrease, which is optimistic for wind energy development.

Using the wind power density from 2020 to 2059 and calculation method of seasonal variation index (Sv), we calculate the multi-year average Sv for the next 40 years. At the same time, using the wind power density from 1979 to 2015, we also calculate the multi-year average Sv for the past near 40 years and compare and analyze the past and future Sv, as shown in Fig. 7.6. The greater the value of the Sv, the more significant seasonal difference in wind energy. On the contrary, the smaller the seasonal difference in wind energy.

Sv: Comparing Figs. 6.5 and 6.6, it is not difficult to find that Sv and Mv share the similar spatial distribution characteristics, but the value of Sv is smaller than Mv. Comparing the future Sv with the past Sv, it is found that the future Sv of the North Indian Ocean is less than the historical status as a whole, and the future Sv of the South China Sea is slightly higher than the historical status. In the South China Sea, the future Sv of most sea areas is basically above 1.2. The center is located in the south-central South China Sea (1.5–1.8). The area range with past Sv above 1.5 is significantly smaller than the future value. In the Bay of Bengal, the future Mv is between 0.9 and 1.8, and the Sv range between 1.5 and 1.8 is very small; while the historical state of the sea area's Sv is basically between 1.2–2.1. In the Arabian Sea, the future Mv is between 1.2 and 2.4, and the sea area above 2.1 is very small. Judging from the historical status, the Sv of most regions is between 1.5–2.7, which is significantly higher than the future value.

6.7   Summary

This chapter aims to design a mid-long term projection scheme for wind energy resource, to provide reference for the long term planning of wind energy utilization. We emphasis that: Whether wind energy climate characteristics analysis or, wind energy short-term forecasting or long-term wind energy projection, it is necessary to pay attention to a set of key indicators of wind energy, systematically including a set of key indicators of wind energy (wind power density, energy availability, energy level occurrences, energy stability, energy storage, etc.). The results of long-term projection of offshore wind energy resource show that,

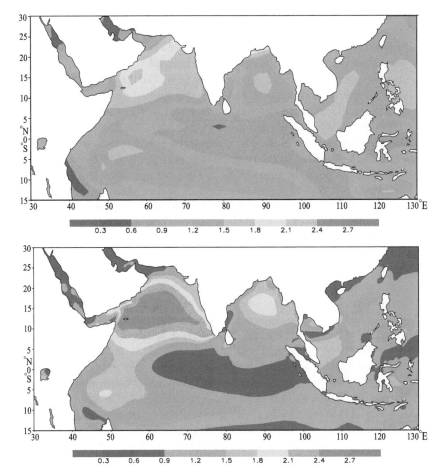

Fig. 6.6 Seasonal variation index of wind power density for the next 40 years (top) and the past near 40 years (bottom)

(1) For wind power density, the projection values along the Maritime Silk Road for the next 40 years show obvious seasonal and regional differences. The wind power density in the Arabian Sea is the largest in August, followed by February and November, and the lowest in May. The Bay of Bengal has the highest in August, followed by May and November, and the lowest in February. The wind power density of the South China Sea in February and November was significantly higher than that in May and August. In terms of annual average value of wind power density, the average wind power density in the South China Sea and the Bay of Bengal for the next 40 years is significantly higher than the historical status, and the projection wind power density in the Arabian Sea for the next 40 years is close to the historical status. The average wind power density in the central and northern regions of the South China Sea for

the next 40 years will basically be above 350 W/m². The average wind power density in the Bay of Bengal for the next 40 years will basically be above 150 W/m², and the center will reach above 350 W/m². The future wind power density in most areas of the Arabian Sea is above 200 W/m².

(2) For the effective wind speed occurrence (EWSO), the seasonal difference of EWSO in the Maritime Silk Road for the next 40 years is reflected in the following: the Arabian Sea has the highest EWSO in August, followed by February and November, and the lowest in May. The Bay of Bengal has the highest in August and November, followed by May, and the lowest in February. The South China Sea has the highest in February, followed by November, and the lowest in May. In terms of the annual average value, the EWSO in the Maritime Silk Road for the next 40 years will be more than 65% (with the center at more than 75%), which is higher than the historical status. The future EWSO in most areas of the central and northern South China Sea is above 70%. The future EWSO in most areas of the Bay of Bengal is above 65%. The average EWSO in the Arabian Sea for the next 40 years is basically above 65%.

(3) For rich level occurrence (RLO), the RLO of the Maritime Silk Road for the next 40 years shows a large seasonal and regional difference. The Arabian Sea has the highest RLO in August, followed by November, and the lowest in May. The Bay of Bengal has the highest in August, followed by November, and the lowest in February. The RLO in the South China Sea is the highest in February, followed by November, and the lowest in May. For the annual value, the RLO along the Maritime Silk Road for the next 40 years is higher than the historical status. The future RLO in most areas of the South China Sea is above 30%. And the area range with RLO above 50% of the future status is wider than that of the historical status. The future RLO in most of the Bay of Bengal is above 20%, and the area range with future RLO above 50% will be wider than the historical status. The future RLO in most of the Arabian Sea is above 40%, and the area range of RLO above 60% of the future status is wider than that of the historical status.

(4) Regarding coefficient of variation (Cv), the spatial distribution of future Cv and past Cv in each month along the Maritime Silk Road remains roughly the same on the whole. The average Cv for the next 40 years is significantly smaller than the historical status. It means that the stability of wind energy for the next 40 years is better than that of the past near 40 years, which is beneficial to resource development. The Cv along the Maritime Silk Road for the next 40 years shows large seasonal differences: August has the best stability in the Arabian Sea, followed by February and November, and May has the worst stability. The Bay of Bengal has the best stability in August, followed by November, and February had the worst stability. The South China Sea has the best stability in February, followed by November, and the worst in May.

(5) Monthly variation index (Mv) of wind energy: On the whole, the spatial distribution of the future Mv and the past Mv are roughly consistent, and the Mv of

the Arabian Sea is significantly larger than the Bay of Bengal and the South China Sea for both past and future. Compared with the historical status of Mv, the projection Mv of the Bay of Bengal and the Arabian Sea tends to decrease, and the projection Mv of the South China Sea is roughly consistent with the historical status, which is optimistic for wind energy development.

(6) Seasonal variation index (Sv) of wind energy: Sv and Mv share roughly the same spatial distribution characteristics, but the value of Sv is smaller than Mv. The future Sv of the North Indian Ocean is generally smaller than the historical status, which indicates that the seasonal difference tends to shrink. The future Sv of the South China Sea is slightly higher than the historical status.

6.8 Prospect

Through the analysis of this chapter, it is also found that: compared with the historical status, the southwest monsoon surge in the Arabian Sea and the Bay of Bengal in the next May will increase in occurrence and weaken in intensity. And the intensity of the southwest monsoon along the Maritime Silk Road in August of the future decades will be significantly lower than the historical status as a whole. In the coming February and November, the cold air along the Maritime Silk Road will show signs of intensification. In the future, it is necessary to conduct an in-depth projection and analysis on the intensity and occurrence of the summer southwest monsoon surge along the South China Sea and North Indian Ocean, as well as the intensity and occurrence of the winter cold airs.

In this chapter, we carry out the medium to long term projection of wave energy along the Maritime Silk Road for the next 40 years, to provide a technology method for the wave energy projection. The study area and time series can be extended in the future according to requirement. For example, the global wave energy for the next 100 years can be projected. In addition, this chapter analyzes the future wind energy along the Maritime Silk Road under the RCP4.5 scenario. In the future, it is necessary to make medium and long-term forecasts of wind energy under different scenarios of RCP2.6, RCP4.5, RCP6.0 and RCP8.5, and compare the similarities and differences of resources under different scenarios, to provide a more detailed reference for the long-term planning of wind energy development.

Regarding the projection of energy level occurrences, this chapter representatively gives the mid- to long-term projection of RLO, which provides a technical way for mid- to long-term projection of different energy level occurrences. It is possible to apply the method of this chapter to carry out mid- and long-term estimates for ALO, MLO, RLO, ELO, and SLO. In this chapter, the CMIP5 data is used. Recently, the CMIP6 data is being launched, which can also be used for wind energy projection.

In this chapter, the CMIP5 data is used in the calculation of future wind energy, and the ERA-Interim reanalysis data is used in the calculation of historical wind energy. The inconsistency of data source may affect the analysis results to a certain extent. In the future, the CMIP6 data can be uniformly used to calculate and compare wind energy resources in the future and historical status, to provide a more accurate scientific basis for the medium and long-term planning of resource development.

In medium and long-term projection, the results are often compared with those of the same era of the last century. This is usually the case for meteorology and oceans. For example, compare the results of the 2080s and 2090s (2080–2099) with the results of the 1980s and 1990s (1980–1999). This method can provide a reference for the mid- and long-term plans, but it has certain shortcomings: (1) It is not very practical. When making medium- and long-term plans, such as resource development, disaster prevention and mitigation, etc., usually more attention is paid to recent decades, but not decades later. If it is estimated in recent decades (for example, 2020–2050), the traditional method is to compare it with the characteristics of 1920–1950. However, there are certain problems such as scarcity of data, processing methods, and data accuracy during the period of 1920–1950, resulting a relative weak value. (2) There is no guarantee that the period of the estimated elements is 100 years. Therefore, it is not necessary to compare the estimated results with the same era of the last century. Taking the current time as the node, comparative analysis of the past decades and the next decades is more practical for the medium and long-term plans.

References

Abolude AT, Zhou W, Akinsanola AA (2020) Evaluation and projections of wind power resources over china for the energy industry using CMIP5 models. Energies 13:2417. https://doi.org/10.3390/en13102417

Capps SB, Zender CS (2009) Global ocean wind power sensitivity to surface layer stability. Geophys Res Lett 36:L09801. https://doi.org/10.1029/2008GL037063

de Winter RC, Sterl A, Ruessink BG (2013) Wind extremes in the North Sea Basin under climate change–an ensemble study of 12 CMIP5 GCMs. J Geophys Res Atmos 118:1601–1612. https://doi.org/10.1002/jgrd.50147

Hirabayashi Y, Mahendran R, Koirala S, Konoshima L, Yamazaki D, Watanabe S, Kim H, Kanae S (2013) Global flood risk under climate change. Nat Clim Chang 3:816–821. https://doi.org/10.1038/nclimate1911

Huntingford C, Jones PD, Livina VN, Lenton TM (2013) No increase in global temperature variability despite changing regional patterns. Nature 500:327–330. https://doi.org/10.1038/nature12310

Li B, Li JM, Li YN (2020) Application of artificial neural network to numerical wave simulation in the coastal region of island. J Xiamen Univer (Nat Sci) 59(3):420–427

Ma DM, Li YS, Liu YX, Cai JW, Zhao R (2021) Vibration deformation monitoring of offshore wind turbines based on GBIR. J Ocean Univer China 20(3):501–511

Poulter B, Frank D, Ciais P, Myneni RB, Andela N, Bi J, Broquet G, Canadell JG, Chevallier F, Liu YY, Running SW, SItch S, van der Werf GR (2014) Contribution of semi-arid ecosystems to interannual variability of the global carbon cycle. Nature 509:600–603. https://doi.org/10.1038/nature13376

Ramesh KV (2014) Assessing reliability of regional climate projections: the case of Indian monsoon. Nat Sci Rep 4:1–9. https://doi.org/10.1038/srep04071

Tambke J, Lange M, Focken U (2005) Forecasting offshore wind speeds above the North Sea. Wind Energy 8:3–16

Taylor KE, Stouffer RJ, Meehl GA (2012) An Overview of CMIP5 and the Experiment Design. Am Meteorol Soc 93:485–498. https://doi.org/10.1175/BAMS-D-11-00094.1

Villarini G, Vecchi GA (2012) In: 21st Century projections of north Atlantic tropical storms from CMIP5 models. Nature 2:604–607.

Wang XL, Feng Y, Swail VR (2014) Changes in global ocean wave heights as projected using multimodel CMIP5 simulations. Geophys Res Lett 41:1026–1034. https://doi.org/10.1002/2013GL058650

Xie S, Lu B, Xiang B (2013) Similar spatial patterns of climate responses to aerosol and greenhouse gas changes. Nat Geosci. https://doi.org/10.1038/ngeo1931

Zheng CW, Pan J, Li JX (2013) Assessing the China Sea wind energy and wave energy resources from 1988 to 2009. Ocean Eng 65:39–48

Zheng CW, Pan J (2014) Assessment of the global ocean wind energy resource. Renew Sustain Energy Rev 33:382–391

Zheng CW, Li XY, Luo X, Chen X, Qian YH, Zhang ZH, Gao ZS, Du ZB, Gao YB, Chen YG (2019) Projection of future global offshore wind energy resources using CMIP Data. Atmos Ocean 57(2):134–148

Zheng CW, Li CY, Pan J, Liu MY, Xia LL (2016) An overview of global ocean wind energy resource evaluations. Renew Sustain Energy Rev 53:1240–1251

Zheng CW, Gao Y, Chen X (2017) Climatic long term trend and prediction of the wind energy resource in the Gwadar Port. Acta Scientiarum Naturalium Universitatis Pekinensis 53(4):617–626

Zheng CW (2018) Wind energy evaluation of the 21st Century Maritime Silk Road. J Harbin Eng Univ 39(1):16–22

Chapter 7
Offshore Wind Energy Evaluation in the Sri Lankan Waters

Sri Lanka is known as a pearl in the Indian Ocean. It is located at the southern end of South Asia. It faces the Indian Peninsula across the Pauk Strait in the north, and is close to the equator in the south. It holds a forward position on the Indian Ocean and overlooks the important waterway between Europe and Asia. At the same time, Sri Lanka is located near the center of the main channel of the Indian Ocean. It is the main way to the Persian Gulf, the Red Sea, and the Strait of Malacca. It is also an important place for the Maritime Silk Road. Its international trade position is alos extremely critical. A good assessment of offshore wind energy of the Sri Lanka will help promote sustainable development and provide support for the construction of the Maritime Silk Road.

An effective energy evaluation must be carried out before carrying out the resource utilization. Only on the basis of detailed resource investigation can orderly development and utilization be achieved. In 2015, Zheng and Li (2015a) took the lead at home and abroad to conduct the wind and wave energy assessments on the South China Sea. On this basis, the characteristics of wind and wave climate of islands reefs in the South China Sea are further analyzed (2015b), which can provide environmental safety guarantee for resource development. In 2016, Zheng et al. (2016) carried out an assessment of the wind energy resources of Gwadar Port in Pakistan and analyzed the temporal and spatial characteristics of a series of key indicators. In 2017, Zheng et al. (2017) analyzed the historical climatic variation and mechanism of Gwadar Port's wind energy resources, and also carried out a mid-long term projection of wind energy. In 2018, Zheng (2018) analyzed the main energy predicament and counter-measures on key points. In 2019, Zheng and Li (2019) discussed the Maritime Silk Road marine new energy assessment from the perspective of marine powers. In their study, they discussed the important role of new marine energy in the construction of a maritime power, the current situation of resource assessment and the main difficul-ties of resource assessment (detailed investigation of energy climate characteristics, energy classification, correlation between energy and important astronomical factors, short-term forecasting of energy, long-term climatic variation of energy, mid- and long-term projection, energy evaluation of key nodes), and provided measures to deal with these difficulties. The above work provides a technical approach for the evaluation of wind energy resources on islands and reefs along the Maritime Silk Road.

On the whole, the current research focusing on wind energy evaluation of the Maritime Silk Road is relatively rare, and the wind energy evaluation on the remote islands and reefs of the Maritime Silk Road is even rarer. Therefore, it cannot provide a good decision support for the wind energy development of key points. Predecessors have done a lot of work on the climatic characteristics, energy classification, climatic variation, and short-term forecasting of wind energy, but there are very few studies on the medium- and long-term projection of wind energy of key points, which are closely related to the medium- and long-term development plan. This chapter takes advance in launching the mid- and long-term planning of wind energy resources for key points, and takes the wind energy projection of Sri Lanka for the next 40 years as an example, to provide scientific and technological support and decision-making support for the mid- and long-term planning of wind energy projects such as offshore wind power and desalination.

7.1 Data and Methods

7.1.1 Data Introduction

ERA-Interim reanalysis: the data introduction is shown in Chap. 3.
    CMIP5 data: the data introduction is shown in Chap. 6.

7.1.2 Method Introduction

The historical 6-hourly wind power density from 00:00 on January 1, 1979 to 18:00 on December 31, 2015 was calculated by using the ERA-Interim reanalysis data. We also calculated the 3-hourly wind power density from 00:00 on January 1, 2020 to 21:00 on December 31, 2059 in Sri Lanka waters by using the CMIP5 data. Then, a comparative analysis of the characteristics of wind energy in Sri Lanka waters for the next 40 years (shorten as future status) and the past near 40 years (shorten as past status) was carried out to provide a scientific basis for the medium and long-term planning of wind energy development. The research content covers a full set of key indicators of wind energy: wind power density, wind energy availability, energy richness, energy stability, the characteristics of energy direction (co-occurrence of wind power density and wind direction), the long-term climatic trend of wind energy, occurrence of wind class, etc.

In this chapter, the CMIP5 data used in the calculation of future wind energy is not uniform with the ERA-Interim re-analysis data used in the calculation of historical wind energy, which will affect the analysis results to some extent. In the future, the CMIP6 data can be uniformly used to calculate and compare wind energy resources in the future and historical status, to provide a more accurate scientific basis for the medium and long-term planning of resource development.

7.2 Temporal-Spatial Distribution of Offshore Wind Energy for the Next 40 years

7.2.1 Wind Power Density

The average of wind power density in January 2020 can be obtained by averaging the wind power density from 00:00 on January 1, 2020 to 21:00 on January 31, 2020. In the same way, the average wind power density in every January of the next 40 years can be obtained. Then the multi-year average wind power density in January for the next 40 years can be got. The same method can be applied to calculate the monthly characteristics of wind power density in the Sri Lanka waters for the next 40 years and the monthly characteristics of wind power density in this region for the past near 40 years, as shown in Fig. 7.1.

On the whole, the monthly variation of wind power density in Sri Lanka waters for the future shares a similarity with that of the past, and both show the monthly variations of one main peak and one sub-peak. The peak appears in June (wind

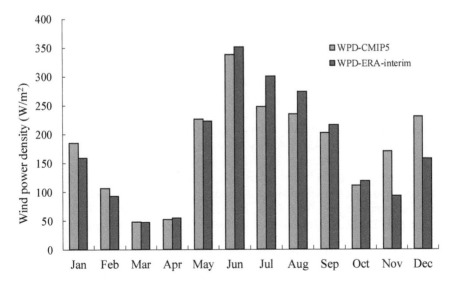

Fig. 7.1 Monthly characteristic of wind power density for the next 40 years and the past near 40 years in the Sri Lanka waters

power density in the future and in the past are both around 350 W/m^2). The sub-peak occurres in December (230 W/m^2 in the future, 160 W/m^2 in the past). And the two troughs appeared in March–April and October–November respectively. During the summer monsoon, the future wind power density is slightly lower than the past value, while the situation is the opposite during the winter monsoon. It shows that under the rcp4.5 scenario, the southwest monsoon in the sea area tends to weaken, while the winter monsoon strengthens. It is generally considered that WPD above 50 W/m^2 is usable, and 200 W/m^2 or above is rich. Obviously, wind energy in the Sri Lanka waters is available all year round, and nearly half of the months of the year are abundant.

Zheng et al. (2016) analyzed the wind energy resources of 10 m above the sea surface of Gwadar Port and found that Gwadar Port contains relatively rich wind energy resources. The annual average wind power density during 1979–2014 was 121 W/m^2. This paper calculates that the average wind power density of Sri Lanka waters during 1979–2014 was 173 W/m^2, and the average wind power density for the next 40 years was 179 W/m^2, which is obviously more abundant than Gwadar Port.

7.2.2 Energy Availability

It is generally believed that the wind speed between 5 and 25 m/s is beneficial to the collection and conversion of wind energy, and the wind speed in this interval is defined as the effective wind speed for wind energy resource development (abbreviated as effective wind speed). Obviously, statistical analysis of the effective wind speed occurrence (EWSO) can show the availability of wind energy. Using the 3-hourly CMIP5 data from 00:00 on January 1, 2020 to 21:00 on January 31, 2020, the statistics of the EWSO in January 2020 is obtained. Similarly, the EWSOs in each January for the next 40 years are obtained. Then the multi-year average EWSO in January for the next 40 years is obtained. Using the same method, the multi-year average EWSOs from January to December for the next 40 years and the near past near 40 years are obtained. In the same way, the monthly variation characteristics of EWSO for the past near 40 years, by using 6-hourly ERA-interim data from the 00:00 on January 1, 1979 to 18:00 on December 31, 2015, as shown in Fig. 7.2.

The EWSO in the Sri Lanka waters have similar monthly variations in the future and past, and both show a monthly variation of one main peak and one sub-peak. The peak occurs in June (both future and past EWSO is above 80%). The sub-peak occurs from December to January of the following year. The two troughs appear in March–April and October–November respectively. The EWSO during the summer monsoon for the future is slightly lower than that for the past, while the situation is the opposite during the winter monsoon. Regardless of the past or the future, the EWSO in the Sri Lanka waters has exceeded 50% for 7 months throughout the year, which is optimistic for wind energy development.

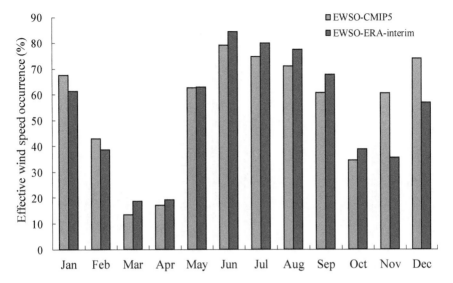

Fig. 7.2 Monthly characteristic of effective wind speed occurrences (EWSO) for the next 40 years and the past near 40 years in the Sri Lanka waters

Zheng et al. found that the annual EWSO of Gwadar Port during 1979–2014 was 43%, which has a higher utilization rate than the solar energy wich is restricted by daylight and bad weather. According to statistics, this chapter found that the annual EWSO in the Sri Lanka waters during 1979–2014 was 53%, and the annual EWSO during 2020–2059 was 55%. The availability of wind energy is significantly higher than that of the Gwadar Port.

7.2.3 Energy Richness

It is generally considered that wind power density above 200 W/m^2 is rich. Obviously, statistical analysis of occurrence of wind power density greater than 200 W/m^2 (rich level occurrence, RLO) can show the degree of enrichment of wind energy. Through the statistics of wind power density from 00:00 on January 1, 2020 to 21:00 on January 31, 2020, we can get the RLO in January 2020 and the monthly RLO in the next 40 years, then the monthly variation characteristics of RLO under the multi-year average status are also obtained. In addition, the monthly variation characteristics of the RLO in the multi-year average state over the past near 40 years can also be obtained, by using the 6-hourly wind power density from 00:00 on January 1, 1979 to 18:00 on December 31, 2015, as shown in Fig. 7.3.

The monthly variation characteristics of RLO and EWSO show certain similarities, but the values of RLO are low. Regardless of the future or the past, the overall RLO of the Sri Lanka waters is optimistic: the RLO for 7 months in the whole year

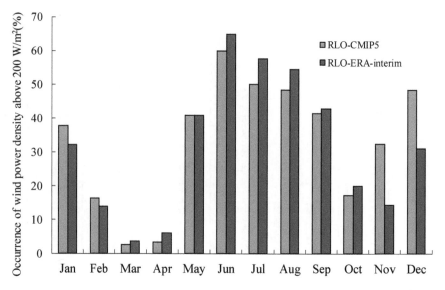

Fig. 7.3 Monthly characteristic of occurrences of wind power density greater than 200 W/m² (rich level occurrence, RLO) for the next 40 years and the past near 40 years in the Sri Lanka waters

is more than 30%. In summer, it is as high as 50%, which shows relatively abundant wind energy resources.

7.2.4   Energy Direction

A stable direction of resources is conducive to improving the collection and conversion efficiency of wind energy, and also helps to extend the life of the wind turbine. On the contrary, it will seriously affect the efficiency of wind energy development, and even affect the life of the wind turbine. This chapter draws a wind energy rose diagram for the next 40 years and the past near 40 years, and uses February, May, August, and November as the representative months of winter, spring, summer and autumn to show the characteristics of wind energy in the Sri Lanka waters, as shown in Fig. 7.4.

In February: Wind energy of the Sri Lanka waters steadily comes from the northeast (NE) direction. For the next 40 years, the occurrence of east-northeast (ENE) direction will be as high as 45%, followed by the NE direction (30%), and the occurrences in the other directions will be lower. The high wind power density above 200 W/m² will mainly come from the NE direction, followed by the ENE direction. For the past near 40 years, the occurrence of NE direction was 33%, followed by the ENE direction (31%); the high wind power density above 200 W/m² mainly originated from the NE direction.

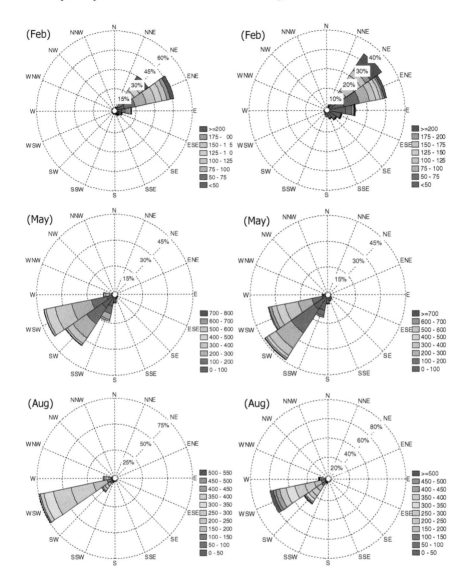

Fig. 7.4 Wind energy rose for the next 40 years (left) and the past near 40 years (right) in the Sri Lanka waters

In May: The wind energy of the Sri Lanka waters mainly comes from the southwest (SW) direction. It is not difficult to find that the sea area has completed the transition from the winter monsoon to the southwest monsoon in May. For the next 40 years, the most occurrence directions will be west-southwest (WSW) and SW, with occurrences of 42% and 35%, respectively. The occurrence of wind power density above 200 W/m² in May is significantly higher than that in February, which

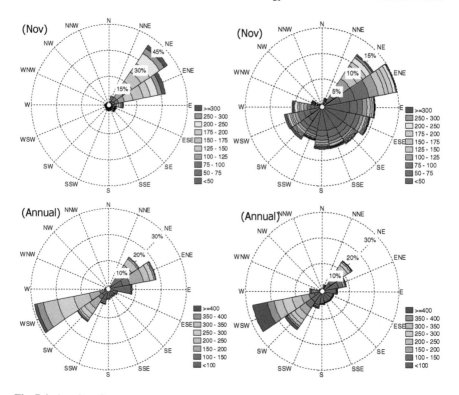

Fig. 7.4 (continued)

is also mainly derived from WSW and SW directions. For the past near 40 years, the most occurrence directions are SW (44%) and WSW (36%). The wind power density above 200 W/m² also mainly originated from these two directions.

In August: the WSW direction occupies an absolute dominant position in wind energy of the Sri Lanka waters. For the next 40 years, the occurrence of the WSW direction will be as high as 75%, and the occurrence of energy levels above 200 W/m² will be above 50%. For the past near 40 years, the occurrence of WSW direction was 60%, followed by SW direction (24%); occurrence above 200 W/m² mainly appeared in WSW direction with occurrence of 55%.

In November: For the next 40 years, NE direction (41%) and ENE direction (33%) will dominate. The wind power density above 200 W/m² will also mainly come from these two directions. For the past near 40 years, energy has mainly come from NE-SE-SW, which is quite different from the next 40 years.

Annual: Whether the next 40 years or the past near 40 years, the wind energy of the Sri Lanka waters is steadily contributed by WSW to SW and NE to ENE, which is very beneficial to wind energy development. For the next 40 years, WSW will have the highest occurrence (close to 30%), followed by ENE (18%), followed by SW and NE. The wind power density above 200 W/m² will mainly appear in WSW, with an occurrence as high as 24%. For the past near 40 years, the occurrence of occurrence of WSW was 28%, followed by SW (18%), followed by NE (14%). The wind power density above 200 W/m^2 mainly appeared in WSW.

7.2.5 Energy Stability

Energy stability is directly related to the efficiency of wind energy collection and conversion. Cornett (2008) once judged the stability of wave energy by calculating the coefficient of variation (Cv) of wave power density. The smaller the Cv, the better the stability. With reference to Cornett's calculation method, this chapter uses the wind power density from January 1, 2020 00:00 to January 31, 2020 21:00 to calculate the Cv of wind energy in January 2020 and Cv in every month for the next 40 years. Then the monthly characteristics of Cv under the multi-year average status for the next 40 years is obtained. Using the same method, with the 6-hourly wind power density from 00:00 on January 1, 1979 to 18:00 on December 31, 2015, to obtain the monthly characteristics of Cv under multi-year average status for the past near 40 years, as shown in Fig. 7.5.

On the whole, regardless of the next 40 years or the past near 40 years, the Cv of wind energy in the Sri Lanka waters will show a "M"-shaped bimodal monthly characteristics. Two crests appear in March and October, and two troughs appear in June–August and December. Obviously, the poor wind energy stability in March and October is due to the poor wind speed stability during the monsoon transition season; while the better wind energy stability in June–August and December is due to the

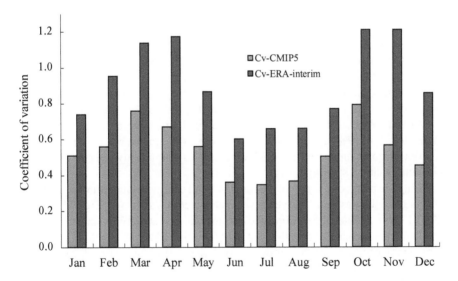

Fig. 7.5 Monthly characteristic of coefficient of variation (Cv) of wind power density for the next 40 years and the past near 40 years in the Sri Lanka waters

strong and regular summer southwest monsoon and winter northeast monsoon. It is not difficult to find that the Cv of the next 40 years is generally smaller than the Cv of the past near 40 years, which means that the stability of wind energy in the next 40 years is better than that of the past near 40 years.

7.3 Future Trend in Wind Energy

The average wind power density in 2020 can be obtained by averaging the 3-hourly wind power density from 2020.01.01 to 2020.12.31. Using the same method, we obtain the average wind power density of the Sri Lanka waters in each year for the next 40 years, and analyze its linear trend, as shown in Fig. 7.6a. By processing the 3-hourly wind speed data from 2020.01.01 to 2020.12.31, the EWSO in 2020 can be obtained. Using the same method, we obtain the annual EWSO in each year for the next 40 years, and analyze its linear trend, as shown in Fig. 7.6b. Using the same method as in Fig. 7.6b, we calculate and analyze the annual trends of RLO and Cv in the Sri Lanka waters for the next 40 years, as shown in Fig. 7.6c, d.

As shown in Fig. 7.6, the linear correlation (|R|) of WPD is 0, failed the significance test. EWSO's |R|=0.11, failed the significance test. RLO's |R|≈0, failed the significance test Test; Cv's |R|=0.08, which failed the significance test; indicating that the WPD, EWSO, RLO, and Cv in the Sri Lanka waters have no significant annual trend for the next 40 years.

Using the method of Fig. 7.6, the trends of WPD, EWSO, RLO, and Cv in February, May, August, November of the Sri Lanka waters for the next 40 years are calculated, as shown in Table 7.1. It is not difficult to find that the various elements of wind energy in the Sri Lanka waters have a significant trend in February and May, but there is no significant trend in August and November. In February, the WPD, EWSO, and RLO decreased significantly at a rate of -0.6525 (W/m²)/yr, -0.3771%/yr, and -0.1635%/yr (the % here refers to EWSO or RLO, not their variation rate, the same below), respectively, which indicates that wind energy is slowly decreasing. The Cv increases at a rate of 0.0033, which indicates that the stability tends to deteriorate. In May, the WPD, EWSO, and RLO increased significantly at the rate of 2.5841 (W/m²)/yr, 0.4419%/yr, and 0.4510%/yr respectively, indicating that wind energy showed a significant increasing trend; Cv increases significantly at a rate of -0.0060, indicating that the stability tends to be better.

Using the method in Fig. 7.6, the trend of wind energy resources in the Sri Lanka waters over the past near 40 years was calculated, as shown in Fig. 7.7. WPD's |R|=0.17, which failed the significance test; indicating that the WPD in Sri Lanka has no significant annual trend in the past near 40 years. EWSO's |R|=0.48, passed the 99% significance test, and the regression coefficient is 0.15, indicating that the EWSO in the Sri Lanka waters has been increasing at a rate of 0.15%/yr (the % here refers to EWSO, not the variation rate of EWSO, the same below) in the past near 40 years. RLO's |R|=0.37, passed the 95% significance test, and the regression coefficient was 0.1063, indicating that the RLO of the Sri Lanka waters has increased significantly

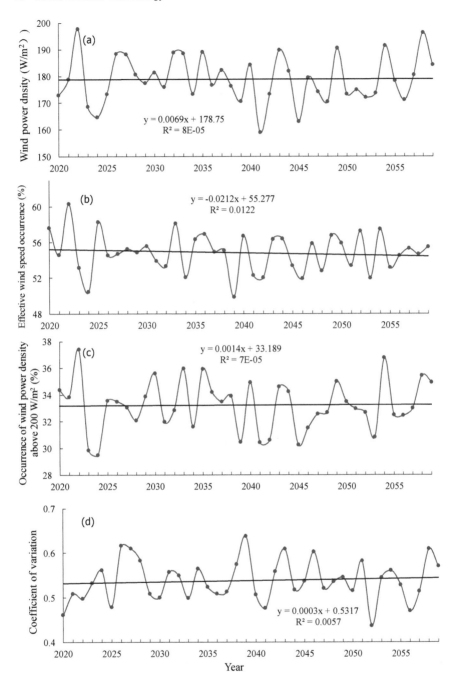

Fig. 7.6 Annual trends of wind power density (a), effective wind speed occurrence (b), occurrences of wind power density greater than 200 W/m^2 (rich level occurrence, RLO, c) and coefficient of variation (Cv) in the Sri Lanka waters for the next 40 years

Table 7.1 Trends of wind energy elements based on CMIP5 in the next 40 years

Time	Wind energy density (W/m^2)/yr	Effective wind speed frequency %/yr	Frequency above 200 W2/m %/yr	Coefficient of Variation
February	−0.6525**	−0.3771***	−0.1635**	0.0033**
May	2.5841****	0.4419****	0.4510****	−0.0060***
August	−0.0755	0.0090	0.0298	−0.0009
November	0.1327	0.0321	−0.0411	0.0009
Year by year	0.0069	−0.0212	0.0014	0.0003

*, **, ***, ****indicate that it has passed the reliability test of 90%, 95%, 99%, and 99.9%, respectively

at a rate of 0.1063%/yr (the % here refers to RLO, not the variation rate of RLO, the same below) over the past near 40 years. Cv's |R|=0.34, which passed the 95% significance test, and the regression coefficient was -0.0017, indicating that the Cv of the Sri Lanka waters has been decreasing at a rate of -0.0017 in the past near 40 years, and its stability has gradually improved. Obviously, in the past near 40 years, the wind energy resources of the Sri Lanka waters have tended to be optimistic: the size and availability of wind energy have increased significantly year by year, and the stability has also tended to improve.

Using the method in Table 7.1, the trends of WPD, EWSO, RLO, and Cv in the Sri Lanka waters for the past near 40 years were calculated, as shown in Table 7.2. It is not difficult to find that the various elements of wind energy of the Sri Lanka waters have a significant variation in February, May and August, but there is no significant variation in November. In February, the WPD, EWSO, and RLO increased significantly at a rate of 1.2703 (W/m^2)/yr, 0.4735%/yr, and 0.3357%/yr, respectively, indicating that the wind energy is increasing. The Cv has no significant change trend. In May, there was no significant trend in WPD. The EWSO and RLO increased significantly at the rate of 0.5573%/yr and 0.4598%/yr, respectively, indicating that wind energy showed a significant increasing trend. The Cv decreases significantly at a speed of -0.0105, indicating that the stability tends to be better. In August, the WPD, EWSO, and RLO decreased significantly at a rate of -1.1756 (W/m^2)/yr, -0.1446%/yr, and -0.2044%/yr, respectively, indicating a significant decline in wind energy. The Cv at a rate of 0.0013 significance increases, indicating that stability tends to deteriorate.

7.4   Summary and Prospect

This chapter uses the CMIP5 data to calculate and analyze the wind energy characteristics of the Sri Lanka waters for the next 40 years and uses the ERA-interim data to calculate the wind energy characteristics of this area for the past near 40 years. In addition, the characteristics of wind energy for the next 40 years and the past near

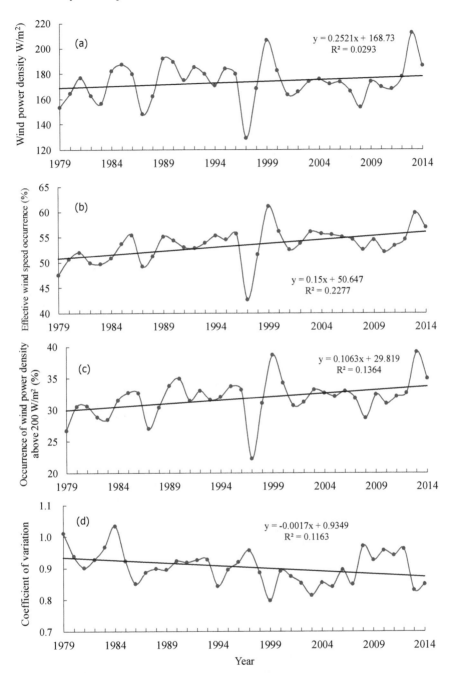

Fig. 7.7 Annual trends of wind power density (a), effective wind speed occurrence (b), occurrences of wind power density greater than 200 W/m^2 (rich level occurrence, RLO, c) and coefficient of variation (Cv) in the Sri Lanka waters for the past near 40 years

Table 7.2 Variation trends of wind energy elements based on ERA-interim in the past near 40 years

Time	Wind energy density (W/m²)/yr	Effective wind speed frequency %/yr	Frequency above 200 W2/m %/yr	Coefficient of Variation
February	1.2703***	0.4735***	0.3357***	−0.0012
May	1.3220	0.5573***	0.4598**	−0.0105****
August	−1.1756**	−0.1446**	−0.2044**	0.0013
November	0.1168	0.0986	0.0503	−0.0018
Year by year	0.2521	0.1500***	0.1063**	−0.0017**

*, **, ***, **** indicate that it has passed the reliability test of 90%, 95%, 99%, and 99.9%, respectively

40 years are also compared and analyzed, to provide scientific reference for the wind energy development of this region. The result is as follows,

(1) Regardless of the future or the past, the wind power density in the Sri Lanka waters has shown monthly characteristics of one main peak and one sub-peak. The peak appeared in June, and the sub-peak appeared in December. The two troughs appeared in March–April and October–November respectively. The annual average wind power density for the next 40 years is 179 W/m², which is higher than the past near 40 years.

(2) The monthly characteristics of EWSO and RLO in the Sri Lanka waters are similar to those of wind power density. The overall performance of EWSO and RLO for the next 40 years is optimistic. The annual averages are 55% and 33%, respectively, which are higher than the past near 40 years. The EWSO for 7 months in the whole year is above 50%.

(3) Whether the next 40 years or the past near 40 years, the stability of wind energy in the Sri Lanka waters will be steadily contributed by WSW to SW and NE to ENE, which is very beneficial to wind energy development. For the next 40 years, the WSW direction has the highest occurrence (close to 30%), followed by the ENE direction (18%). The wind power density above 200 W/m² mainly appears in the WSW direction, with an occurrence as high as 24%. For the past near 40 years, the occurrence of WSW was 28%, followed by SW (18%). The wind power density above 200 W/m² mainly appeared in WSW direction.

(4) The Cv of wind energy in the Sri Lanka waters all show an "M"-shaped bimodal monthly variation, that is, the stability during the winter monsoon and summer monsoon is significantly better than that during the monsoon transition season, which is the same for the next 40 years and the past near 40 years. The Cv in the next 40 years is generally smaller than the Cv in the past near 40 years, which means that the stability of wind energy in the next 40 years is better than that in the past near 40 years.

(5) For the next 40 years, the WPD, EWSO, RLO, and Cv in the Sri Lanka waters will not change significantly year by year. For the past near 40 years, the wind

energy resources in the Sri Lanka waters have tended to be optimistic: the size and availability of wind energy have been significant year by year increasingly, the stability also tends to improve. Whether in the next 40 years or the past near 40 years, the changing trends of various elements of wind energy show significant seasonal differences.

This chapter evaluates the wind energy resources in the Sri Lanka waters for the next 40 years, and compares and analyzes the temporal and spatial distribution characteristics of various elements of wind energy for the next 40 years (future status) and the past near 40 years (historical status), which can provide a scientific basis for the medium and long-term planning of wind energy development. By referring to this plan, in the future, medium and long-term forecasts can be made for offshore wind energy resources in more remote islands and reefs.

In addition, this chapter systematically analyzes the characteristics of various elements of wind energy for the next 40 years. In the future, referring to the energy classification scheme constructed by Zheng and Li (2018), which comprehensively considers resource characteristics, marine environment, and cost-effectiveness, to carry out classification of wind energy resources for the next 40 years (or 10 years in the future). At the same time, combining and comparing the classifications of historical and future wind energy to better provide decision-making support for wind energy development site selection, and to facilitate the orderly and efficient deployment of offshore wind power, seawater desalination and other wind energy projects.

In this chapter, the data used in the calculation of future wind energy is CMIP5, while the data used in the calculation of historical wind energy is ERA-Interim reanalysis. Therefore, the inconsistency of data will affect the analysis results to a certain extent. In the future, the CMIP6 data can be uniformly used to calculate and compare wind energy resources in the future and historical status, to provide a more accurate scientific basis for the medium and long-term planning of resource development.

References

Zheng CW, Li CY (2015) Development of the islands and reefs in the South China Sea: Wind power and wave power generation. Periodical Ocean Univ China 45(9):7–14

Zheng CW, Li CY (2015) Development of the islands and reefs in the South China Sea: Wind climate and wave climate analysis. Periodical Ocean Univ China 45(9):1–6

Zheng CW, Li CY, Yang Y, Chen X (2016) Analysis of wind energy resource in the Pakistan's Gwadar Port. J Xiamen Univ (Nat Sci Edn) 55(2):210–215

Zheng CW, Gao Y, Chen X (2017) Climatic long term trend and prediction of the wind energy resource in the Gwadar Port. Acta Scientiarum Naturalium Universitatis Pekinensis 53(4):617–626

Zheng CW (2018) Energy predicament and countermeasure of key junctions of the 21st Century Maritime Silk Road. Pacific J 26(7):71–78

Zheng CW, Li CY (2019) Evaluation of new marine energy for the Maritime Silk Road from the perspective of maritime power. J Harbin Eng Univ 41(2):175–183

Chapter 8
Construction of Temporal-Spatial Characteristics Dataset of Offshore Wind Energy Resource

The power supply capacity along the road is weak, which severely restricts the efficient development of the Maritime Silk Road. On the whole, the electricity consumption along the "One Belt and One Road" is only 61% of the world average level (Jiang et al. 2019). How to break the power dilemma has become the core of the efficient development of the Maritime Silk Road. The offshore wind energy, with advantages of renewable, pollution-free, all-weather and many other advantages, has become a new focus pursued by developed countries. It mainly used for offshore wind power, seawater desalination, etc. Predecessors made a great contribution to the offshore wind energy evaluation in global oceans (Carvalho et al. 2014; Zheng et al. 2016a, 2016b; Capps and Zender 2010). But so far, data on offshore wind energy resources are scarce, which is the key basis for achieving efficient development of wind energy.

The production and sharing of marine data has become an important manifestation of comprehensive national strength. Compared with land-based data, it is more difficult to obtain the marine data. This dilemma significantly affects the efficient development of related marine development and construction (Yang et al. 2021; Zheng and Li 2017). Currently, the marine and meteorological raw data are relatively abundant. However, the offshore wind energy data is scarce. How to extracted the useful information about wind energy development from the original big data with large volume and low information density, and then building a wind energy resource dataset has become the key support to the industrialization and efficient deployment of wind energy. It is also a common challenge for global colleague.

This study aims to establish the first open-ended and non-profit dataset of spatial–temporal characteristics of wind energy for the Maritime Silk Road at home and abroad, to provide scientific reference and data support for offshore wind energy resource evaluation, thus to overcome the power dilemma, thus to contribute to the construction of the Maritime Silk Road.

C. Zheng et al., *21st Century Maritime Silk Road: Wind Energy Resource Evaluation*, Springer Oceanography, https://doi.org/10.1007/978-981-16-4111-4_8

8.1   Data and Methods

In this study, the first open-ended and nonprofit temporal-spatial dataset of offshore wind energy resource for the 21st Century Maritime Silk Road was established (Zheng 2020). First, we systemically presented the calculation methods of a series of key parameters of wind energy. Then we extracted a range of useful information about wind energy development from the original big data with large volume and low information density, systematically including the wind power density (WPD), effective wind speed occurrence (EWSO), energy level occurrences, coefficient of variation (Cv), monthly variability index (Mv), seasonal variability index (Sv), total storage, effective storage and technological storage. This dataset can provide data support for relevant decision makers, researchers and engineers, and serve as scientific basis for the efficient industrialization of offshore wind energy development, as well as a technical way to overcome the power dilemma faced by the construction of the Maritime Silk Road.

The initial wind data used in this study is the ERA-interim data that hosted at the ECMWF. The data introduction of the ERA-Interim reanalysis is shown in Chap. 3.

Since 2012, Zheng et al. (2012, 2019) has proposed the effective wind speed occurrence (EWSO) and energy level occurrences, to describe the availability and richness of wind energy, and continued to improve these parameters. These two parameters are widely recognized and used by domestic and foreign colleagues. The dataset proposed by this study systematically covers a series of key indicators such as WPD, EWSO, energy level occurrences, Cv, Mv, Sv, and energy storage.

8.2   Calculation Methods of Parameters

Wind power density (WPD). It is defined as the mean annual power available per square meter of swept area of a turbine, with calculation as Eq. (3.1). Using the ERA-interim wind data, according to the calculation method of WPD, the 6-hourly WPD for the past near 40 years is calculated. Averaging the WPD from 00:00 January 01, 1979 to 18:00 January 31, 1979, the monthly average WPD in January 1979 at each $0.25° \times 0.25°$ bin is obtained. Similarly, the monthly average WPD in each January for the past near 40 years is obtained. Then the data of multi-year average January WPD at each $0.25° \times 0.25°$ bin is calculated. In the same way, the data of multi-year average WPD from January to December are obtained. This data file is named "WPD.dat" and contains 12 intervals from January to December. The spatial range is 30°E–130°E, 15°S–30°N, spatial resolution: $0.25° \times 0.25°$. The missing value is -9.99e + 33. The temporal-spatial distribution characteristics of WPD along the Maritime Silk Road are presented in Sect. 3.2.

Effective wind speed occurrence (EWSO). During the exploitation, it is recognized that the wind speed between 5 to 25 m/s is defined as the EWSO, the optimal speed for collection and transformation of wind power. EWSO could also be defined as the occurrence of wind speed between 5 to 25 m/s. Generally, offshore wind power is more abundant than that on land, and hence the former standard is adopted hereby. The equation of wind power availability is as Eq. (3.2). Based on the 6-hourly wind speed data from 00:00 January 01, 1979 to 18:00 January 31, 1979, the EWSO in January 1979 at each 0.25° × 0.25° bin is counted. Similarly, the EWSO in each January for the past near 40 years is counted. Then the data of multi-year average January EWSO at each 0.25° × 0.25° bin is calculated. In the same way, the data of multi-year average EWSO from January to December are obtained. This data file is named "EWSO.dat" and contains 12 intervals from January to December. The spatial range of the data: 30°E–130°E, 15°S–30°N, spatial resolution: 0.25° × 0.25°. The missing value is -9.99e + 33. The temporal-spatial distribution characteristics of EWSO along the Maritime Silk Road are presented in Sect. 3.3.

Energy level occurrences. Since 2011, Zheng and Li (2011) groundbreakingly proposed the wave energy level occurrences, to depict the richness of wave power. Zheng et al. (2012) proposed the wind energy level occurrences to depict the wind power richness. In the early stage, this index mainly included the available level occurrence (ALO, occurrence of wind power density greater than 100 W/m^2) and rich level occurrence (RLO, occurrence of wind power density greater than 200 W/m^2). This study improves and further develops the wind energy level occurrences and set five level standards: ALO, moderate level occurrence (MLO, occurrence of WPD greater than 150 W/m^2), RLO, excellent level occurrence (ELO, occurrence of WPD greater than 300 W/m^2), superb level occurrence (SLO, occurrence of WPD greater than 400 W/m^2, as shown in Table 8.1. The equation of energy level occurrences is as Eqs. (3.3–3.7). Based on the 6-hourly WPD data from 00:00 January 01, 1979 to 18:00 January 31, 1979, the ALO in January 1979 at each 0.25° × 0.25° bin is counted. Similarly, the ALO in each January for the past near 40 years is counted. Then the data of multi-year average January ALO at each 0.25° × 0.25° bin is calculated. In the same way, the data of multi-year average ALO from January to December are obtained, as well as the values of MLO, RLO, ELO, SLO. The temporal-spatial distribution characteristics of energy level occurrences along the Maritime Silk Road are presented in Sect. 3.4.

Coefficient of variation. It demonstrates the stability of the WPD in each season or month. The smaller the Cv is, the more stable the density. The calculation method of Cv is as Eqs. (3.8). Using the 6-hourly WPD data from 00:00 January 01, 1979 to 18:00 January 31, the Cv in January 1979 is calculated. Similarly, the Cv in each January for the past near 40 years is obtained. Then the data of multi-year average January Cv at each 0.25° × 0.25° bin is calculated. In the same way, the data of multi-year average Cv from January to December are obtained. This data file is named "Cv.dat" and contains 12 intervals from January to December. The spatial range of the data: 30°E–130°E, 15°S–30°N, spatial resolution: 0.25° × 0.25°. The missing value is -9.99e + 33. The temporal-spatial distribution characteristics of Cv along the Maritime Silk Road are presented in Sect. 3.5.

Monthly variability index. It demonstrates the monthly differences of wind power. The larger the value of Mv is, the more variable the monthly difference of wind power and vice versa. The calculation method of Mv is as Eqs. (3.10). Averaging the WPD from 00:00 January 01, 1979 to 18:00 January 31, 1979, the monthly average WPD in January 1979 at each $0.25° \times 0.25°$ bin is obtained. Similarly, the monthly average WPD in each month of 1979 is obtained. Then the Mv in 1979 is calculated according to the calculation method of Mv. In the same way, the Mv of each year for the past near 40 years is obtained. Then the multi-year average Mv for the past near 40 years is calculated. This data file is named "Mv.dat" and contains 1 interval. The spatial range of the data: 30°E–130°E, 15°S–30°N, spatial resolution: $0.25° \times 0.25°$. The missing value is -9.99e + 33. The temporal-spatial distribution characteristics of Mv along the Maritime Silk Road are presented in Sect. 3.6.

Seasonal variability index. It demonstrates the seasonal differences of wind power. The larger the value of Sv is, the more variable the seasonal difference of wind power and vice versa. The calculation method of Sv is as Eqs. (3.11). Averaging the WPD from 00:00 March 01, 1979 to 18:00 May 31, 1979, the seasonal average WPD in March–April-May (MAM) 1979 at each $0.25° \times 0.25°$ bin is obtained. Similarly, the seasonal average WPD in each season of 1979 is obtained. Then the Sv in 1979 is calculated according to the calculation method of Sv. In the same way, the Sv of each year for the past near 40 years is obtained. Then the multi-year average Sv for the past near 40 years is calculated. This data file is named "Sv.dat" and contains 1 interval. The spatial range of the data: 30°E–130°E, 15°S–30°N, spatial resolution: $0.25° \times 0.25°$. The missing value is -9.99e + 33. The temporal-spatial distribution characteristics of Sv along the Maritime Silk Road are presented in Sect. 3.6.

Energy storage. It is closely linked to the volume of electricity generated. Previous researchers have tremendous work in this perspective but mostly on the storage in total. With reference to the method proposed by Zheng et al. (2016), we calculated the wind power storage, including the total storage, exploitable storage and technological storage on each $0.25° \times 0.25°$ grid point. The equations are as Eqs. (3.12–3.14). Based on the WPD and EWSO data for the past 40 years, the wind energy storage on each $0.25° \times 0.25°$ grid point is calculated. This data file is named "Storage.dat" and contains three parameters: total storage, exploitable storage and technological storage. The spatial range of the data: 30°E–130°E, 15°S–30°N, spatial resolution: $0.25° \times 0.25°$. The missing value is -9.99e + 33. The temporal-spatial distribution characteristics of energy storage along the Maritime Silk Road are presented in Sect. 3.7.

8.3 Value and Significance

The value of this dataset lies in the establishment of the first set of wind energy resource data for the Maritime Silk Road. The useful information for wind energy evaluation and development is extracted from the original data with large volume and low information density. This dataset systematically covers a series of key indicators that wind energy evaluation and development pay close attention to, including the multi-year average WPD from January to December, multi-year average EWSO from January to December, multi-year average ALO from January to December, multi-year average RLO from January to December, multi-year average Cv from January to December, the multi-year average Mv, the multi-year average Sv, and wind energy storage (including the total storage, exploitable storage and technological storage). This dataset is a key support for the efficient and scientific development of wind energy projects such as offshore wind power generation and seawater desalination, thus making positive contributions to overcoming the energy dilemma faced by the Maritime Silk Road. In addition, the wind energy resource dataset established in this study is a finished product, which is more conducive to shortening the construction period of related projects than the raw marine data.

8.4   Usage Notes and Recommendations

The temporal-spatial characteristics dataset of offshore wind energy resource for the 21st Century Maritime Silk Road has been uploaded to Science Data Bank for release (available at http://www.dx.doi.org/10.11922/sciencedb.j00001.00142). At the same time, users are required to add citations and mark "Zheng Chong-wei Team of Dalian Naval Academy" in the acknowledgment. This is an open-ended and non-profit dataset, which is only open to the public welfare construction and scientific research, and not open to any commercial organization or department.

8.5   Prospect

In this Chapter, the temporal-spatial characteristics dataset of offshore wind energy resource was established by extracting the useful information about wind energy evaluation from the ERA-interim data. In addition, the ERA5 data is available online now, which includes the wind data of 100 m height above the sea surface. In the future, it is necessary to establish the wind energy resource dataset of 100 m height above the sea surface, according to the method of this study.

In this Chapter, the temporal-spatial characteristics dataset of offshore wind energy resource was established. In the future work, a wind energy resource, which should include the temporal-spatial characteristics, energy classification, climatic variation of wind energy, long-term projection of wind energy, wind climate, etc., can be established to provide a systematic data support for power plant site selection, daily operation, long-term planning, environmental safety guarantee for resource development.

Table 8.1 Dataset profile (Zheng 2020)

Title	Temporal-spatial characteristics dataset of offshore wind energy resource for the 21st Century Maritime Silk Road
Data author	Zheng Chongwei
Data corresponding author	Zheng Chongwei (chinaoceanzcw@sina.cn)
Time range	Multi-year average values from January to December
Data volume	16 MB
Geographical scope	30°E–130°E, 15°S–30°N
Data format	*.dat
Data service system	< http://www.dx.doi.org/10.11922/sciencedb.j00001.00142 >
Sources of funding	Major International (Regional) Joint Research Project of National Science Foundation of China (41,520,104,008)
Dataset composition	This dataset is composed of 8 files of binary data as follows: (1) Monthly wind power density (WPD) of multi-year average status of the Maritime Silk Road from January to December, named "WPD.dat" (2) Monthly effective wind speed occurrence (EWSO) of multi-year average status of the Maritime Silk Road from January to December, named "EWSO.dat" (3) Monthly available level occurrence (ALO, occurrence of WPD greater than 100 W/m^2) of multi-year average status of the Maritime Silk Road from January to December, named "ALO.dat" (4) Monthly rich level occurrence (RLO, occurrence of WPD greater than 200 W/m^2) of multi-year average status of the Maritime Silk Road from January to December, named "RLO.dat" (5) Coefficient of variation (Cv) of multi-year average status of the Maritime Silk Road from January to December, named "Cv.dat" (6) Monthly variability index (Mv) of multi-year average status of the Maritime Silk Road from January to December, named "Mv.dat" (7) Seasonal variability index (Sv) of multi-year average status of the Maritime Silk Road from January to December, named "Sv.dat" (8) Total storage, effective storage and technological storage of wind energy of the Maritime Silk Road from January to December, named "Storage.dat"

References

Capps SB, Zender CS (2010) Estimated global ocean wind power potential from QuikSCAT observations, accounting for turbine characteristics and siting. J Geophys Res 115:D09101. https://doi.org/10.1029/2009JD012679

Carvalho D, Rocha A, Gómez-Gesteira M et al (2014) Offshore wind energy resource simulation forced by different reanalyses: Comparison with observed data in the Iberian Peninsul. Appl Energy 134:57–64

Dee DP, Uppala SM, Simmons AJ et al (2011) The ERA-Interim reanalysis: configuration and performance of the data assimilation system. Q J R Meteorol Soc 137(656):553–597

Huang QE, Xu JJ (2020) One-year experiment and evaluation of atmospheric reanalysis in East Asia and nearby coastal zones. J Xiamen Univ Natl Sci 59(3):401–411

Jiang Y, Wu MQ, Huang CJ et al (2019) Collection of data on overseas power projects in the belt and road initiative (2000–2019). China Sci Data 4(4):(2019–12–28). https://doi.org/10.11922/csdata.2019.0069.zh

Liu HW, Wang WB, Jiang Y et al (2016) Advantages and limitations of reanalysis data in wind energy resource assessment. Wind Energy 12:58–63

Song LN, Liu ZL, Wang F (2015) Comparison of wind data from ERA-Interim and buoys in the Yellow and East China Seas. Chin J Oceanol Limnol 33(1):282–288

Yang SB, Xi LT, Li XF, Zheng CW (2021) Temporal and spatial characteristics of wave energy resources in Sri Lankan waters over the past 30 years. J Ocean Univer China 20(3):489–500

Zheng CW, Li CY, Yang Y et al (2016) Analysis of Wind Energy Resource in the Pakistan's Gwadar Port. J Xiamen Univ (Nat Sci) 55(2):210–215

Zheng CW, Li CY (2017) 21st Century Maritime Silk Road: Big Data Construction of New Marine Resources: Wave Energy as a Case Study. Ocean Dev Manage 34(12):61–65

Zheng CW, Zhuang H, Li X et al (2012) Wind energy and wave energy resources assessment in the East China Sea and South China Sea. SCIENCE CHINA Technol Sci 55(1):163–173

Zheng CW, Li XY, Luo X et al (2019) Projection of Future Global Offshore Wind Energy Resources using CMIP Data. Atmos Ocean 57(2):134–148

Zheng CW, Li XQ (2011) Wave Energy Resources Assessment in the China Sea During the Last 22 Years by Using WAVEWATCH-III Wave Model. Periodical Ocean Univ China 41(11):5–12

Zheng CW, Li CY, Pan J et al (2016) An overview of global ocean wind energy resource evaluations. Renew Sustain Energy Rev 53:1240–1251

Zheng CW. Temporal-spatial characteristics dataset of offshore wind energy resource for the 21st Century Maritime Silk Road. China Sci Data 5(4):(2020–12–28). https://doi.org/10.11922/csdata.2020.0097.zh

Lightning Source UK Ltd.
Milton Keynes UK
UKHW021108150822
407314UK00002B/9